PROTEIN ENGINEERING

Titles published in the series:

*Antigen-presenting Cells

*Complement

Enzyme Kinetics

Gene Structure and Transcription

Genetic Engineering

*Immune Recognition

*B Lymphocytes

*Lymphokines

Membrane Structure and Function

Molecular Basis of Inherited Disease

Protein Engineering

Regulation of Enzyme Activity

*Published in association with the British Society for Immunology.

Series editors

David Rickwood
Department of Biology, University of Essex, Wivenhoe Park, Colchester, Essex CO4 3SQ, UK

David Male
Institute of Psychiatry, De Crespigny Park, Denmark Hill, London SE5 8AF, UK

PROTEIN ENGINEERING

Peter C.E. Moody
and
Anthony J. Wilkinson
Department of Chemistry, University of York

at
OXFORD UNIVERSITY PRESS
Oxford New York Tokyo

Oxford University Press
Walton Street, Oxford OX2 6DP

Oxford is a trade mark of Oxford University Press

Published in the United States
by Oxford University Press, New York

© Oxford University Press, 1990

All rights reserved. No part of this publication may be reproduced stored in a retrieval system, or transmitted in any form by any means, electronic, mechanical, photocopying, recording, or otherwise, without the prior permission of Oxford University Press

This book is sold subject to the condition that it shall not, by way of trade or otherwise, be lent, re-sold, hired out, or otherwise circulated without the publisher's prior consent in any form of binding or cover other than that in which it is published and without a similar condition including this condition being imposed on the subsequent purchaser

British Library Cataloguing in Publication Data
Moody,P.C.E.
  Protein engineering.
  1. Protein engineering
  I. Title   II. Wilkinson,A.J.
  547.75
  ISBN 0-19-963194-8

Library of Congress Cataloging in Publication Data
Moody, Peter C.E.
  Protein engineering / Peter C.E.Moody and Anthony J.Wilkinson.
    p.   cm.—(In focus)
  Includes bibliographical references (p.   ) and index.
  1. Proteins—Biotechnology.   I. Wilkinson, Tony   II. Title.
  III. Series: In focus (Oxford, England)
  TP248.65.P76W55 1990   660'.63—dc20   90-7728
  ISBN 0-19-963194-8

Typeset and printed by Information Press Ltd, Oxford, England.

Preface

This book sets out to provide an introduction to, and a description of, protein engineering; a substantial and flourishing area of biochemistry. The conceptual origin and the definition of protein engineering are considered in Chapter 1. Protein engineering is very much an interdisciplinary field, with substantial contributions from protein chemistry, enzymology, protein crystallography, and molecular genetics. In a book of this size it is not possible to survey individually these subjects with any rigour; instead, in Chapters 2 and 3, we discuss the aspects of these areas of direct relevance to protein engineering. In Chapters 4 and 5, we have tried to emphasize the enormous importance and potential of protein engineering both as a tool for exploring how proteins work and as a route to designing proteins for industrial and medical uses. These aspects are extensively discussed in terms of examples taken from recent scientific literature. In this way we hope to illustrate the scope and limitations of our current understanding of proteins, as well as to convey some of the excitement in the field at present. This book is suitable not only for undergraduates studying chemistry or biological sciences who are interested in proteins, but it will also serve as a useful introductory text to research workers moving into this field.

Finally, we would like to thank all our colleagues at York for their support. In particular, we gratefully acknowledge Marek Brzozowski, Guy Dodson, Mike Hartshorn, Madeleine Moore, Steve Smerdon, Chandra Verma, Dave Edwards, Paul Emsley, and Rod Hubbard variously for providing illustrations, helpful comments, and careful reading of the manuscript.

P.C.E.Moody
A.J.Wilkinson

Contents

Abbreviations	xi
1. Introduction	1
Historical perspective	1
Protein evolution	2
Protein engineering	3
References	3
2. Protein structure	5
Introduction	5
Amino acids and proteins	5
The main chain	5
Properties of side chains	9
Protein structural organization	11
Examination of protein structure	12
Structure determination	12
Assessing model quality	15
Analysis of mutant structures	16
Further reading	17
References	17
3. Preparation and analysis of mutant proteins	19
Introduction	19
Expression systems	19
Overview	19
The *lac* system	20
The λ P_L system	20
Recombinant DNA methods	22
Overview	22
Restriction endonucleases	22

DNA ligase	22
Plasmids	23
Bacterial transformation	24
Site-directed mutagenesis	25
Oligonucleotide-directed mutagenesis	25
Screening for mutants	25
Increasing the yield of mutants	27
Analysis of mutant proteins	28
Overview	28
Protein–ligand interactions	28
Enzyme kinetics	29
The energetics of enzyme catalysis	31
Further reading	33
References	34

4. Protein engineering: site-directed mutagenesis as a probe of function — 35

Introduction	35
Tyrosyl-tRNA synthetase	35
Reaction and properties	35
Probing the role of cysteine-35 in enzyme–substrate binding	36
Hydrogen-bonding and specificity	39
Catalysis of tyrosyl-adenylate formation	42
The serine proteases	44
Introduction	44
Mechanism of action	45
Expressing serine protease mutants	47
The catalytic triad	48
Transition-state stabilization	49
Chloramphenicol acetyl transferase	50
Myoglobin: discrimination between O_2 and CO	52
Structure and properties	52
Role of the distal histidine in myoglobin	52
Repressor proteins	55
Introduction	55
Recognition helices	56
Helix swapping	57
Conclusions	59
Further reading	59
References	59

5. Tailoring protein properties and function — 61

- Introduction — 61
- Engineering faster-acting insulins — 61
  - Insulin and diabetes — 61
  - Problems of therapeutic insulins — 62
  - Protein engineering of monomeric insulins — 63
- Antibodies — 66
  - The molecular immune system — 66
  - Immunoglobin structure — 66
  - Antibody therapy — 67
  - Reshaping human antibodies — 68
  - Catalytic antibodies — 68
- Enhancing activity in tyrosyl-tRNA synthetase — 69
- Subtilisin: practical applications — 71
  - Overview — 71
  - Engineering resistance to chemical oxidation — 71
  - Modification of the pH/activity profile — 72
  - Altering substrate specificity — 73
- Engineering thermostability in lysozyme — 74
  - Protein stability — 74
  - Conformational entropy — 76
  - α-Helix dipole interactions — 76
  - Introducing covalent crosslinks — 77
- Conclusions — 78
- Further reading — 78
- References — 78

Glossary — 81

Index — 83

Abbreviations

Ala	alanine
Asn	asparagine
Asp	aspartic acid
ATP	adenosine triphosphate
CAT	chloramphenicol acetyl transferase
Cm	chloramphenicol
CoA	coenzyme A
Cys	cysteine
DFP	di-isopropyl fluorophosphate
DNA	deoxyribonucleic acid
dNTP	deoxyribonucleotide triphosphate
Gln	glutamine
Glu	glutamic acid
Gly	glycine
His	histidine
Ile	isoleucine
Leu	leucine
αLP	α-lytic protease
Lys	lysine
Mb	myoglobin
mRNA	messenger RNA
NAD^+	nicotinamide adenine dinucleotide
NMR	nuclear magnetic resonance
Phe	phenylalanine
PMSF	phenylmethylsulphonyl fluoride
RNA	ribonucleic acid
RNAp	RNA polymerase
SBT	subtilisin
SDS	sodium dodecyl sulphate
Ser	serine
TIM	triose phosphate isomerase
Thr	threonine
TPCK	tosyl-phenylalanyl chloromethyl ketone
tRNA	transfer RNA
Trp	tryptophan
Tyr	tyrosine
TyrRS	tyrosyl-tRNA synthetase
Val	valine

1

Introduction

1. Historical perspective

Protein engineering represents the culmination of a series of spectacular advances in molecular biology in the second half of this century. It is salutary to recall that DNA was only recognized as the hereditary material in 1944. In the ensuing years, landmarks in molecular biology have included the determination of the double helical structure of DNA in the 1950s, the elucidation of the genetic code in the 1960s and the development of techniques for recombining and sequencing DNA in the 1970s.

In parallel, dramatic breakthroughs were being made in protein structure determination by X-ray analysis of protein crystals. The three-dimensional atomic structures of proteins revealed immediately how they were constructed and explained their general behaviour in solution. For the first time the stereochemistry at their active sites and surfaces was defined, allowing chemists to visualize the spatial relationships of the active residues whose chemistry they had been investigating so laboriously. The importance of these structures cannot be overstated, since they provide a solid framework both for understanding function and formulating mechanisms of action.

The facility for extending and exploiting our understanding of protein structure has been an achievement of the 1980s. It has emerged through a marriage of molecular genetics and crystallography as protein engineering. Protein engineering is built on the central principle of molecular evolution—that new properties and functions emerge from existing structures through the occurrence and accumulation of spontaneous mutations in the gene. Evolution is formidable and inexorable; it is also blind and sluggish. As described in this book, it is now possible to introduce mutations at precise positions within a gene using *in vitro* techniques, bypassing the natural evolutionary mechanisms to accelerate and to redirect these processes. This has provided a powerful capacity for redesigning proteins at will.

2. Protein evolution

Proteins play leading and very diverse roles in biology; in catalysis, the enzymes; in immunity, the antibodies; in growth and development, the DNA-binding proteins; in the transport of metabolites, the carrier proteins; in the relaying of biological signals, the hormones; and in biological structures, the fibrous proteins. The diversity of structures and functions that are associated with proteins is the product of millions of years of evolution. Mutations arise randomly and spontaneously in proteins through mutations in the DNA. It is selective pressures on the organism that provide the direction. A mutation that confers an advantage to the organism will persist and form the new wild-type. In contrast, mutations that disadvantage the organism will be lost.

Evolution in action is amply illustrated by studies of the amidase gene from the bacterium *Pseudomonas aeruginosa* (1, 2). The aliphatic amidase hydrolyses amides to their corresponding acids and ammonia according to the following reaction:

$$R-CONH_2 + H_2O \rightarrow R-COO^- + NH_4^+$$

The activity of the enzyme is restricted to substrates where R is a methyl (in acetamide) or an ethyl (in propionamide) group. Expression of the amidase gene enables *P. aeruginosa* to grow on agar plates supplemented with acetamide as the only source of carbon and nitrogen for biosynthesis. Amidases with altered substrate specificities have been evolved in the laboratory by applying directed selective pressure on the amidase gene. In initial experiments, cultures of the bacteria were spread on to agar plates in which butyramide replaces acetamide as the sole source of carbon and nitrogen. Butyramide with its extra ethyl group relative to acetamide is a very poor substrate for the wild-type enzyme. Of the cells plated out, only approximately 1 in 10^6 are able to grow and form colonies. Careful genetic and biochemical analysis of these 'adapted' cells showed that in a number of cases, the amidase enzyme had increased activity towards butyramide. Subsequently, cells expressing the mutant amidase were plated on selective media where phenylacetamide was the sole source of carbon and nitrogen. From cells that grew under these conditions, an 'amidase' with activity towards phenylacetamide that possessed no detectable activity towards acetamide was produced, thus demonstrating that the amidase system has the genetic potential for considerable evolutionary change.

In this work, it was possible to direct genetic pressure on a specific gene to achieve an alteration in the characteristics of the encoded enzyme. Unfortunately, it is not possible to use this approach generally because loss or alteration of the protein may not be selectable. Furthermore, such a selective approach is suitable in the laboratory only for micro-organisms for which it is possible to screen large populations of cells for rare mutational events.

3. Protein engineering

Protein engineering is concerned with the construction, analysis, and uses of modified proteins. The aim is to tailor protein properties and activities in a predetermined way. This may include changing the substrate specificity of an enzyme or increasing the stability of a protein for industrial or therapeutic uses.

Protein engineering has emerged from the application of the methods of molecular genetics and genetic engineering to problems in protein chemistry and enzymology. Many of the most interesting proteins are rare in nature and their purification in the quantities necessary for biochemical assay requires prohibitive amounts of biological tissue as starting material. This problem is overcome if the gene can be isolated and the protein expressed in cell cultures such as yeast or *Escherichia coli*. It is now possible to introduce any desired mutation into a cloned and sequenced segment of DNA using the technique of oligonucleotide-directed mutagenesis. This allows amino acid sequences in the encoded proteins to be rearranged or substituted at will. When combined with a detailed knowledge of the structure and biochemistry of the protein, this technique is a powerful probe of mechanism of action and provides the potential for 'adapting' proteins rationally.

It has been pointed out that protein engineering is not engineering in the sense that the word is used in civil engineering (3). For instance, one would not attempt to build a suspension bridge if our understanding of mechanics and material science was as imprecise as our understanding of the ground rules for macromolecular interactions. Nevertheless, it can be argued that imposing architectural structures were built in ancient times (e.g. Roman aqueducts) largely on the basis of intuition rather than on the basis of sound engineering principles. In protein engineering experiments, the starting point is usually precisely defined—a three-dimensional structure of a protein from crystallographic studies and a cloned and sequenced gene. Similarly, desired modifications can be achieved with great precision using site-directed mutagenesis of the gene. In contrast, it is possible to define the end-product only in terms of a desired function. This is because there is not a full enough understanding as to how structure determines function or how the amino acid sequence of a protein determines its three-dimensional structure. The operational definition of protein engineering is expanded because of these constraints, to encompass studies of modified proteins in general. This is a fertile area at present and is allowing protein–ligand interactions and protein stability to be dissected into individual components and contributions. The experience gained and the measurements made in these studies are progressively providing the fuller understanding that will, and already is, enabling real engineering of proteins.

4. References

1. Betz,J.L., Brown,P.R., Smyth,M.J. and Clarke,P.H. (1974) *Nature*, **247**, 261.
2. Clarke,P.H. (1980) *Proc. R. Soc. Lond. B,* **207**, 385.
3. Knowles,J.R. (1987) *Science,* **236**, 1252.

2

Protein structure

1. Introduction

Proteins are extremely complex structures, a small protein (myoglobin) is shown in *Figure 2.1*. The figure is difficult to extract useful information from, solely because of this complexity. It is therefore necessary to examine structures piece by piece. Fortunately, in engineering proteins, it is often necessary to consider only small regions of the protein, in this case it might be the oxygen-binding pocket as discussed in Section 5 of Chapter 4 and shown in *Figure 4.8*. None the less, it is essential to understand the chemistry of proteins and the ways in which their structures are determined in order to be able to modify them rationally.

2. Amino acids and proteins

2.1 The main chain

Proteins are polymers formed by condensation of L-α amino acids [H_2NCH(R) COOH] to make peptide bonds. The protein main chain is formed by a backbone of peptide bonds and α-carbon atoms to which the side-chains (R) are attached.

$$\cdots\text{N}-\underset{\underset{\text{H}}{|}}{\overset{\overset{\text{R}}{|}}{\text{C}}}-\overset{\overset{\text{O}}{\|}}{\text{C}}-\text{N}-\underset{\underset{\text{H}}{|}}{\overset{\overset{\text{R}}{|}}{\text{C}}}-\overset{\overset{\text{O}}{\|}}{\text{C}}-\text{N}-\underset{\underset{\text{H}}{|}}{\overset{\overset{\text{R}}{|}}{\text{C}}}-\overset{\overset{\text{O}}{\|}}{\text{C}}-\text{N}-\underset{\underset{\text{H}}{|}}{\overset{\overset{\text{R}}{|}}{\text{C}}}-\overset{\overset{\text{O}}{\|}}{\text{C}}\cdots$$

There are twenty different amino acids that are incorporated into polypeptide chains, giving variety to the monotonous repeat of the backbone. The structures of the side chains are simple chemical groups and are illustrated in *Figure 2.2*; and their physical properties are summarized in *Table 2.1*. The primary structure

6 Protein Engineering

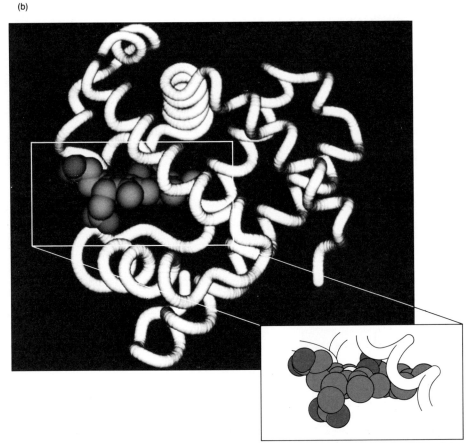

Figure 2.1. The myoglobin molecule: all non-hydrogen atoms are drawn in *Figure 2.1a*, showing the complexity of this small protein. In *2.1b* the main chain is traced, showing the folding of the protein and the position of the bound haem group.

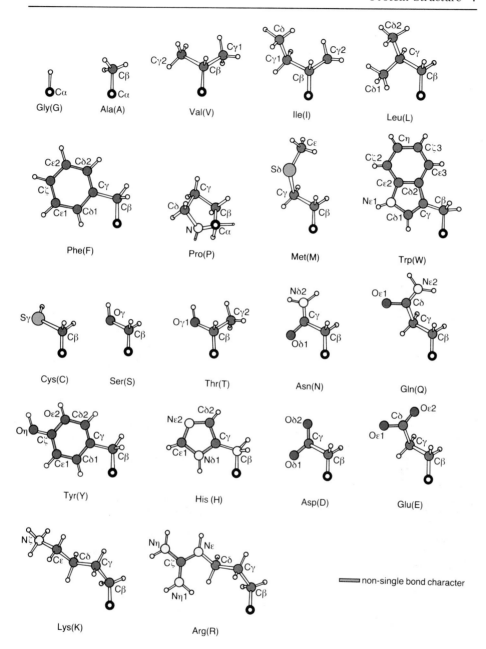

Figure 2.2. The amino acid side-chain groups, their three-letter codes, and single-letter abbreviations. (After Shultz G.E. and Schirmer R.H. (1979))

of a protein is defined as the linear sequence of its amino acid residues, and by convention the order from left to right is from the amino end (N-terminus) to the carboxyl end (C-terminus). This is the order in which the constituent amino acids are assembled on the ribosome.

Table 2.1. The properties of the amino acid residues

Amino acid residue	Code	pK_a of side chain	ΔG for transfer of side chain from water to ethanol (kcal/mol at 25°C)	Area (Å^2) in standard state[a]
Alanine	Ala		−0.5	118.1
Glutamate	Glu	4.07		186.2
Glutamine	Gln			193.2
Aspartate	Asp	3.9		158.7
Asparagine	Asn			165.5
Leucine	Leu		−1.8	193.1
Glycine	Gly			88.1
Lysine	Lys	10.79		225.8
Serine	Ser		+0.3	129.8
Valine	Val		−1.5	164.5
Arginine	Arg	12.48		256.0
Threonine	Thr		−0.4	152.5
Proline	Pro			146.8
Isoleucine	Ile			181.0
Methionine	Met		−1.3	203.4
Phenylalanine	Phe		−2.5	222.8
Tyrosine	Tyr	10.13	−2.3	236.8
Cysteine	Cys	8.35		146.1
Tryptophan	Trp		−3.4	226.3
Histidine	His	6.04	−0.5	202.5

[a] Area for a given residue, X, the standard state accessibility is defined as the average surface area that residue has in a representative ensemble of Gly-X-Gly tripeptides.
Data are from references 2 (pK_as), 3 (area), and 4 (hydrophobicity).

The chemical nature of the peptide bonds puts constraints on the conformation of the main chain. In the late 1930s Pauling and Corey determined experimentally that a planar group is formed from one α-carbon atom to the next, and deduced that this is because the peptide bond has significant double-bond character. In this planar group, the HN–CO groups are usually (with the exception of proline) *trans* (*Figure 2.3*). The main-chain amide (>NH) and carbonyl (>C=O) groups can act as hydrogen-bond donors and acceptors, respectively. An important consequence of this is that hydrogen bonds NH ··· O=C can be formed, allowing segments of the protein main chain to link together.

It can be seen that the planes of the peptide bonds are connected by the N–Cα and Cα–C bonds, whose rotations are described by the dihedral angles ϕ and ψ, respectively (*Figure 2.3*). These angles are mutually dependent, as they are constrained by the position of adjacent peptide bond; free rotation is further limited by the Cβ substituent. The allowed combinations are illustrated in the Ramachandran (1) plot of ϕ against ψ for residues with Cβ substituents (*Figure 2.4b*) and for glycine which has only a hydrogen-atom side chain (*Figure 2.4a*). Glycine has greater conformational freedom, as the broad range of allowed ϕ and ψ angles attest. In contrast, proline, which is an imino acid, has its NH group locked in the pyrrolidine ring, which constrains ϕ to be −60°.

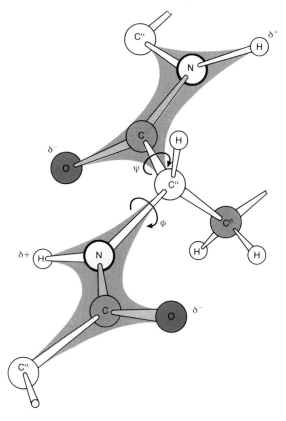

Figure 2.3. The planarity of the peptide bond. Partial double bonds are shown in colour. The dihedral angles ψ and ϕ are indicated, and the dipole is shown by the partial charges $\delta+$ and $\delta-$.

2.2 Properties of side chains

In this section the notable chemical and physical properties of the amino acid side chains are considered, and their contributions to the variety of form and function found in proteins are outlined. The potential of protein engineering stems from the facility for substituting one of these groups for another. The side chain groups are shown in *Figure 2.2*, and some of their most important chemical and physical properties are summarized in *Table 2.1*.

Alanine, valine, leucine, and isoleucine have aliphatic groups that vary in their size and thus their hydrophobicity. A direct consequence of this is that they tend to be found in the interior of protein structures, where the tightly packed environment allows the maximum energetic gain from dispersion forces and the minimum disturbance of the solvent. Serine and threonine have small aliphatic side chains with a hydroxyl group capable of making hydrogen bonds. These hydroxyl groups can be involved in interactions with ligands, but their reactivity is limited at neutral pH. However, the reactivity of hydroxyl groups can be greatly

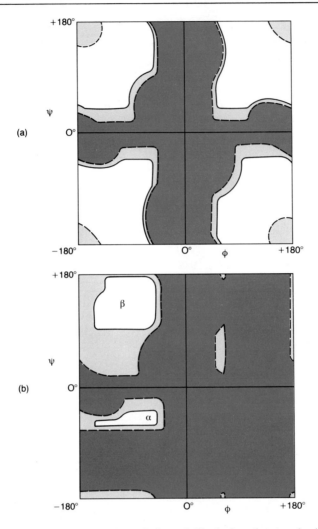

Figure 2.4. Ramachandran plots of allowed dihedral angles ψ and ϕ in polypeptides. The colour shows the regions disallowed by steric clashes if the atoms are considered as hard spheres, the pale colour shows the regions that are allowed by the closest approach distances found experimentally. *2.4a* is for glycine and *2.4b* for peptides with a Cβ substituent. The dihedral angles for the α-helix are shown by the symbol α, and those for a β-strand by β.

increased by the presence of an adjacent base. Cysteine is an intrinsically more reactive side chain. Its pK_a value shows that it is ionized in slightly alkaline conditions. Cysteine residues are also capable of forming disulphides with each other, uniquely allowing covalent bridging of one main chain (or part of a main chain) with another in an oxidizing environment. Methionine also contains sulphur, it is rather hydrophobic; the sulphur atom is strongly nucleophilic and is susceptible to oxidation.

Phenylalanine, tyrosine, and tryptophan have large aromatic side chains with significant absorbances in the ultraviolet part of the electromagnetic spectrum. They are hydrophobic, though tyrosine and tryptophan are able to make hydrogen bonds. Aspartate and glutamate have acidic side chains with pK_a values of around 4, and are therefore extremely polar. Glutamine and asparagine represent their corresponding amides, these polar side chains do not ionize and they can be both hydrogen-bond donors and acceptors. Lysine and arginine have long basic side chains that are also capable of making hydrogen bonds. Unlike the amides, the amino group is normally ionized. Histidine is also capable of ionizing: its pK_a value of 6.0 is close to physiological pH, and it is frequently directly involved in catalysis (see Chapter 4). In its non-ionized form the imidazole ring may exist as two tautomers, one of the nitrogens is an electrophile and hydrogen-bond donor, whilst the other is a nucleophile and hydrogen-bond acceptor. The preferred tautomer will be influenced by the environment of the side chain in the protein. As described previously, glycine and proline are most notable for the effects their side chains have on the main-chain conformation.

When buried inside the protein, charged groups are found with other charged groups of the opposite sign and form ion-pairs. Polar groups are also found at the surface of soluble proteins where they can satisfy their hydrogen-bonding needs with water molecules. When found at the surface of proteins, charged polar groups have further long-range effects due to the electrostatic field generated by the charge. These forces and considerations also apply to the binding of ligands to proteins.

3. Protein structural organization

Examination of the proteins whose three-dimensional structures have been determined show that a remarkable amount of the main chain is formed into regular features of hydrogen-bonding. The local main-chain conformation in these regions is referred to as secondary structure. The significant elements of secondary structure are the α-helix and the β-pleated sheet. The α-helix is a right-handed structure with the backbone carbonyl of each residue hydrogen bonding to the main chain NH of the fourth residue ahead in the chain; this is shown in *Figure 2.5*. Much of the myoglobin structure is α-helix; this can be seen in *Figure 2.1b* by looking at the main chain trace. β-sheets are made up of twisted planes of main chain running either parallel or anti-parallel to each other, with the strands hydrogen-bonded from amide to carbonyl alternately along the sheet. The β-strand is a staggered, extended structure. Inspection of the X-ray structure of proteins has shown that there are often similarities in the organization of secondary structure between proteins, suggesting a common ancestral origin for these proteins. Two well-known examples of these are the nucleotide binding fold found in dehydrogenases and in tyrosyl-tRNA synthetase (see Section 2 of Chapter 4), and the TIM barrel (named after Triose phosphate IsoMerase, the first example) of an eight-stranded β-barrel surrounded by eight helices. Both of these structures are shown in *Figure 2.6*.

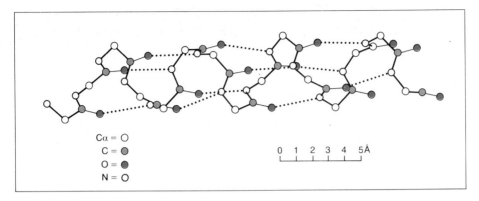

Figure 2.5. The α-helix; hydrogen bonds are shown by dotted lines.

4. Examination of protein structure

4.1 Structure determination

Although the forces that contribute to protein structure and to the binding of ligands are well-understood, it is not possible to predict the three-dimensional structure of a protein from sequence information alone. X-ray crystallography currently provides the only way of determining the full three-dimensional structure of proteins at approaching atomic resolution. However, developments during the 1980s in nuclear magnetic resonance (NMR) spectroscopy have made it possible to determine structures of some small proteins in solution. These experiments do not provide complete structural information but they can provide information on the dynamics of the protein molecule.

A detailed description of protein crystallography would be inappropriate here. None the less, because crystal structures underpin protein engineering, it is important to understand the scope and the limitations of the technique.

Protein crystals are a prerequisite. These crystals should be single and well-ordered with dimensions of the order of tenths of a millimetre. In order to achieve crystallization of a protein, it may be necessary to set up hundreds of crystallization trials, requiring tens or even hundreds of milligrams of pure protein.

The wavelengths of X-rays used for protein crystallography (0.8–1.6 Å, typically) are close to the lengths of covalent bonds. X-ray waves are scattered by the electrons of the atoms in the molecule. In a crystal this scattering is amplified and concentrated in defined directions, producing discrete reflections.

In order to calculate the electron density it is necessary to know both the amplitude and phase of each of a three-dimensional array of reflections. Amplitude information can be measured experimentally, for example, from an image on photographic film (shown in *Figure 2.7*), but phase information cannot be obtained directly. However, it is often possible to modify the protein molecules

Protein Structure 13

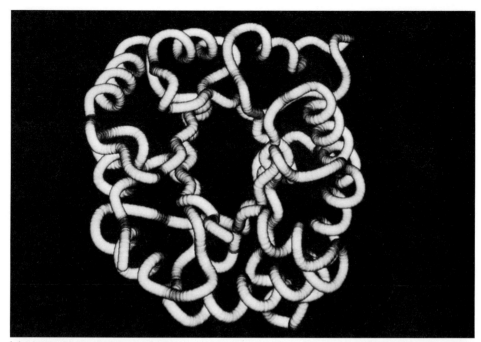

(a)

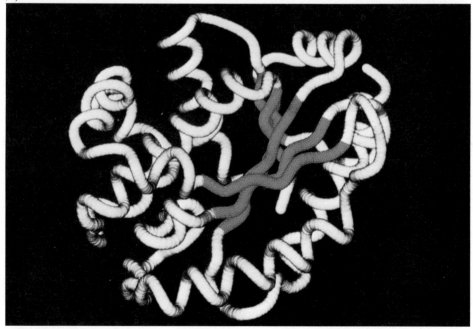

(b)

Figure 2.6. Main chain traces for (a) triose phosphate isomerase looking down the β-barrel and (b) tyrosyl-tRNA synthetase showing the β-sheet in colour.

in the crystal by the inclusion of electron-dense heavy atoms. The electron-dense heavy atoms in the derivative crystals alter significantly the strengths of reflections. Using the differences in the intensities of the native and derivative crystals it is possible to determine the positions of the heavy atoms within the repeating unit of the crystal. If data from more than one such heavy atom derivative are measured, and the quality of data is good, this information can be used to give an estimate of the phase of each reflection in the native crystals. Due to inherent ambiguities and lack of accuracies in the data, it is desirable to have several heavy atom derivatives.

Electron density can be calculated from structure factors (the amplitudes and phases) by applying a Fourier transform to the data. The electron density within the unit cell can be contoured and displayed on a computer graphics device. The known polypeptide sequence can then be built interactively into the electron-density cage or 'chicken wire' (shown in *Figure 2.8*). Stereoscopic viewers make this task considerably easier. The protein model that has been built is then refined by least squares procedures that try to minimize differences between the observed structure factors and those calculated from a back Fourier transform performed on the electron density calculated from the atomic positions of the model. Stereochemical knowledge is used to impose restraints upon the model during refinement. The model can be improved by iterative refinement and model building.

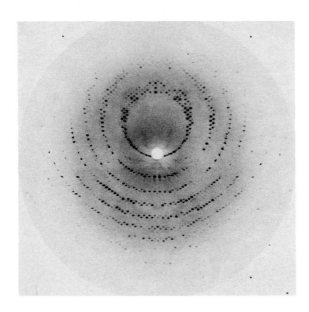

Figure 2.7. An X-ray diffraction photograph from a mutant myoglobin crystal. The crystal has been rotated through 2° during this exposure. About 45 similar photographs need to be taken to collect a full data-set.

4.2 Assessing model quality

It is essential to understand that the atomic co-ordinates produced by protein crystallography represent an interpretation of the observed electron density, albeit a refined interpretation. The newcomer should beware of considering a protein as a fixed set of atomic parameters and take into account the positional accuracy of the model and the movements associated with the atomic positions. The accuracy of a protein model should thus be evaluated critically. The resolution (expressed in Ångstrom units, Å, equal to 0.1 nm) reflects the sampling of the electron density map that can be calculated from the data, that is the resolution determines the sharpness of the electron-density map and thus the detail discernible. The limitations of interpretation at various resolutions are given in *Table 2.2*.

A useful index of the accuracy of a model structure is the *R*-factor, $R = \Sigma(|F_o|-|F_c|)/\Sigma|F_o|$, where $|F_o|$ and $|F_c|$ are the observed and calculated structure factor amplitudes respectively. The lower the *R*-factor, the better the agreement between the observed and calculated structure factors. A typical *R*-value for a well-refined crystal structure at 2.0 Å resolution is less than 0.20. Unrealistically

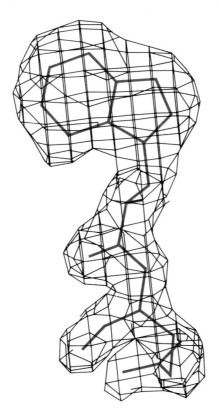

Figure 2.8. The electron density of a tryptophan side-chain observed by X-ray diffraction.

Table 2.2. The limits of interpretation of protein structures determined by X-ray crystallography at different resolutions

Resolution (Å)	Interpretable features
5.5	Helices visible, overall shape of protein.
3.5	Main chain visible, start to see side chains.
3.0	Main chain can be traced with confidence, side chains mostly visible.
2.5	Ordered side chains and plane of peptides resolved, atomic positions assigned to +/− 0.4 Å.
2.0	Hydrogen bonds can be assigned and solvent structure interpreted. Atomic positions to +/− 0.2 Å.
1.5	Individual atoms located to within 0.1 Å.
0.77	Theoretical limit for 1.54 Å radiation, positions to +/− 0.05 Å.

Å = Ångstrom.

lower R-values can be obtained if bond lengths and angles are allowed to deviate from ideal values. It is also possible to improve an R-value by limiting the reflections used in its calculation, direct comparison of these values should not be made without consulting the fine print! The mobility associated with an atom (its B value) is refined at the same time as its co-ordinates. This is equal to $8\pi^2 <u^2>$, where $<u^2>$ is the mean displacement of the atom from its equilibrium position. The average B value for all the atoms in a protein will be about 20, atoms with a B factor of greater than 79 have a root mean square displacement of more than 1.0 Å, and can safely be described as disordered.

An accurately refined model is an object of beauty that is rich in information. It provokes a consideration of the relationship of structure to function. The model is, however, essentially static; it is an average structure of some 10^{15} molecules observed over a period (usually several hours) far greater than that of a biochemical reaction (usually less than a millisecond). The active site and the catalytic mechanism will be of foremost interest if enzymes are being considered. Biochemical data will often be available that help to identify the active site residues, or indeed a substrate or a substrate analogue may have been soaked into the crystals, clearly identifying protein–ligand interactions.

Modern computer graphics allow the structure to be examined closely and substrate models can be built into the active site of an enzyme. This allows extrapolations from the structure to be made which lead to predictions on the intermediate stages of the reaction, and the identification of residues that may be important for catalysis. The conclusions from such modelling exercises are of little value in themselves, but with the facility for testing predictions afforded by altering the structure using site-directed mutagenesis, there is an exquisite approach to analysing mechanisms (Chapter 4).

Predictions of the likely consequences of an amino acid replacement experiment can be made by modelling the possible conformations of the modified side chain.

4.3 Analysis of mutant structures

It is clearly important in interpreting the effects of interesting mutant proteins to determine their structure. This is most easily achieved if it is possible to grow mutant protein crystals that are isomorphous with wild-type crystals. An electron density map may then be calculated using intensities measured from the mutant crystal and phase information from the wild-type structure.

To complement this, the difference Fourier technique may be used to calculate a map from the intensity differences between mutant and wild-type data. New atoms will appear as peaks and the loss of atoms will be shown by troughs in electron density. Shifts in atomic positions also show up as adjacent regions of positive and negative electron density. This provides a guide for building in substituted side chains to make up the mutant model. Subsequent model improvement may be made by refinement.

5. Further reading

Creighton,T.E. (1984) *Proteins*. Freeman, New York.
Shultz,G.E. and Schirmer,R.H. (1979) *Principles of protein structure*. Springer-Verlag, New York.
Blundell,T.L. and Johnson,L.N. (1976) *Protein crystallography*. Academic Press, London.
C.-I.Branden and T.A.Jones (1990) *Nature,* **343**, 647.

6. References

1. Ramachandran,G.N. and Sasisekharan,V. (1968) *Adv. Protein Chem.,* **23**, 283.
2. Dawson,R.M.C., Elliot,D.C., Elliot,W.H. and Jones,K.M. (1969) *Data for Biochemical Research*. Oxford University Press, Oxford.
3. Lesser,G.J., Lee,R.H., Zehfus,M.H. and Rose,G.D. (1987) In Oxender,D.L. and Fox,C.F. (eds), *Protein Engineering*. Alan R. Liss, New York, p. 175.
4. Nozaki,Y. and Tanford,C. (1971) *J. Biol. Chem.,* **246**, 2211.

3

Preparation and analysis of mutant proteins

1. Introduction

Genetic engineering makes it possible to manipulate foreign genes for expression in bacteria. This involves recombining the linear DNA sequence of the gene of interest with sequences that control and initiate gene expression in *E. coli*. Furthermore, it is a routine procedure to introduce any desired mutation into a gene and hence into the protein that it encodes. Here, aspects of these remarkable advances in molecular genetics that provide the tools of protein engineering are discussed before considering approaches to characterizing mutant proteins.

2. Expression systems

2.1 Overview

The amino acid sequence of a protein is enciphered in the base sequence of the DNA of the gene. The linear nucleotide sequence is decoded through the consecutive processes of transcription and translation. With only a few notable exceptions, the genetic code has been conserved throughout evolution. It is therefore possible for eukaryotic genes to be expressed in bacteria. However, the mechanisms and components of the transcription and translation machinery (e.g. RNA polymerases and the ribosomes) are quite different in eukaryotes and prokaryotes. Correspondingly, the sequences in the DNA and mRNA, adjacent to the coding sequence, that interact with these components also differ. Therefore, for efficient expression of a eukaryotic gene in a bacterium such as *E. coli* it is first necessary to recombine the coding sequence of the gene with control sequences that the bacterium can recognize.

The expression of the majority of bacterial genes is controlled at the level of the initiation of transcription (1). Transcription is initiated through the interaction of RNA polymerase with a sequence of DNA upstream of the transcription start site referred to as the promoter. In *E. coli* there is a single RNA polymerase which is responsible for the transcription of all the genes. The level of transcription of a particular gene is determined by the actual sequence of its promoter region.

For protein engineering, the two characteristics that it is desirable to have associated with expression of the gene of interest are, first, high levels of expression—to optimize the yield of protein per cell under given culture conditions, and second, transcriptional control. The latter is needed because overexpression of a foreign protein may well be toxic to the bacterial host, it is advantageous to suppress expression while the cells grow and divide and turn on the expression only after the cells have grown to high density. Two widely exploited promoters are those of the *lac* operon P_{lac}, and of the leftward promoter P_L of bacteriophage λ.

2.2 The lac system

The transcription of the genes of the *lac* operon, which codes for three proteins involved in lactose catabolism (β-galactosidase, lactose permease and β-galactoside transacetylase), is regulated according to the availability of lactose and the need for using lactose as a source of carbon and energy. When lactose is absent from the medium of growing *E. coli* cells there are only five or so molecules of β-galactosidase present in each cell. If lactose is added to the medium, within a short space of time there will be a thousand times as many molecules of β-galactosidase per cell. Expression of the *lac* genes is said to have been induced, and lactose is said to act as an inducer. The mechanism of this induction is now understood in molecular detail.

In the absence of lactose, the genes of the *lac* operon are turned off through the binding of a protein called the *lac* repressor to a sequence of DNA, the operator O_{lac}, that overlaps the promoter region of the DNA. Repressor binds to the *lac* operator with high affinity and high specificity and blocks the binding of RNA polymerase to the promoter and thus the initiation of transcription (*Figure 3.1a*). This repression is relieved in the presence of lactose because lactose, or more precisely an isomer of lactose, allolactose, binds to the *lac* repressor in such a way that repressor can no longer bind to the operator. This allows RNA polymerase to bind to P_{lac} unhindered and initiate transcription (*Figure 3.1b*).

If the coding sequence of the gene of interest is spliced downstream of the P_{lac}/O_{lac} sequences, we can confer the *lac* regulatory properties on that gene. Addition of inducer to cells carrying the DNA sequences of *Figure 3.1c*, will result in high expression of the gene.

2.3 The λ P_L system

A second, widely used expression system, is based on the action of another repressor protein, the bacteriophage (phage) λ repressor which is encoded by the *cI* gene. The phage λ repressor plays an important role in the life cycle of the phage through its tight binding to two operators called O_L and O_R that

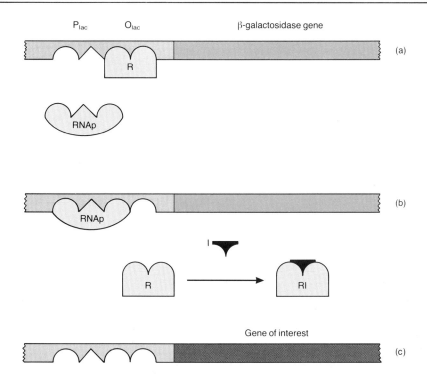

Figure 3.1. Regulation of transcription in the *lac* operon. Repressor (R) binds to O_{lac} (a) preventing the binding of RNA polymerase (RNAp) to the promoter and the initiation of transcription. In the presence of lactose (b), repressor associates with the inducer (I), allolactose. Binding of inducer causes a conformational change in repressor so that its affinity for operator is lowered. RNA polymerase can bind to the promoter unhindered and initiate transcription. (c) When the gene of interest is placed downstream of these regulatory sequences, its expression becomes inducible.

blocks transcription from the λ P_L and λ P_R promoters and maintains the lysogenic (dormant) state of the bacteriophage infection (see Chapter 4, Section 6). Relief of repression of phage λ occurs when repressor protein is inactivated through proteolytic cleavage by the *rec*A protein. *Rec*A protein is produced by the bacterium as a part of the response to DNA damage or ultraviolet irradiation. Cleavage of the repressor leads to expression of genes regulated by P_L and P_R ultimately leading to bacteriophage replication and host-cell lysis (1).

For use as a general expression system, *E. coli* strains have been created that carry a temperature-sensitive λ repressor, referred to as cI_{857}. This protein behaves as a fully functional repressor in cells growing at 30°C but is inactive if the growth temperature is raised to 42°C. Thus, if the coding sequence of the gene of interest is placed downstream of the phage λ P_L promoter, expression of the gene can be induced simply through shifting the growth temperature from 30°C to 42°C. Again, the levels of expression that can be obtained with this promoter are very high. However, repression by cI is tighter than that of *lac* repressor which can be advantageous if small amounts of expression prove lethal to the host cells.

3. Recombinant DNA methods

3.1 Overview
The techniques and strategies for isolating and cloning a gene are outside the scope of this book though they have been succinctly described in the book of this series entitled *Genetic Engineering* (2). This section concentrates on those aspects of genetic engineering relevant to the production of foreign proteins in *E. coli*.

A route to overexpression of a gene in bacteria is to replace the natural promoter of the gene with sequences such as P_{lac} or P_L to exploit the advantageous properties of these promoters. This requires a capacity to cleave DNA sequences specifically and subsequently to splice together heterologous fragments *in vitro*. Nature has provided the necessary tools for recombining DNA *in vitro* in the form of restriction endonucleases and DNA ligase and purified preparations of these enzymes are commercially available.

3.2 Restriction endonucleases
The restriction endonucleases are bacterial enzymes that degrade foreign DNA. They were discovered through their association with the resistance of certain bacteria to infection with bacteriophage. The type II restriction endonucleases recognize DNA sequences 4–6 nucleotide pairs in length and make double-stranded DNA breaks at these sites, leaving 5'-phosphate and 3'-hydroxyl groups at the termini (*Figure 3.2*). The bacterial DNA is itself protected from cleavage through the action of a second enzyme which methylates these sequences so that they are no longer recognized by the restriction endonuclease. The sites of cleavage on each strand may be opposite one another or staggered by one or more base-pairs, but for a given enzyme the positions of cleavage are always the same. Hundreds of restriction enzymes from many sources with various cleavage-recognition sequences are now available.

Digestion of a DNA molecule with a purified restriction endonuclease generates discrete fragments of DNA. Furthermore, if the nucleotide sequence of the DNA that is to be cleaved is known, it is possible to predict the sizes of the fragments that are produced and the sequences that they contain.

3.3 DNA ligase
Fragments of DNA produced by a restriction enzyme digest, with either blunt ends or with complementary cohesive ends, may be joined together by DNA ligase from bacteriophage T4. The enzyme requires ATP as a cofactor and the reaction is favoured by low temperature (4°C).

DNA ligase, the restriction endonucleases and a repertoire of other enzyme activities isolated from various sources together provide a set of tools for

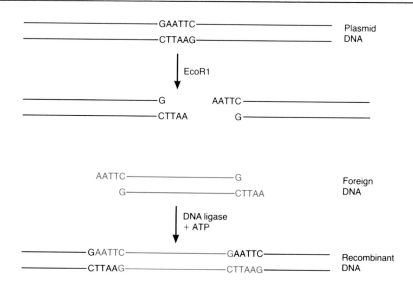

Figure 3.2. Digestion of DNA with the restriction endonuclease *Eco*RI (which recognizes the sequence GAATTC) produces discrete fragments with identical 5' recessive termini. Two *Eco*RI digestion products from different origins (e.g. a plasmid DNA and DNA from the digestion of a chromosome) may be joined end-to-end through the action of DNA ligase to produce a recombinant DNA molecule.

recombining DNA sequences *in vitro* at will. To realize the potential of these DNA constructs, it is necessary to put the recombinant DNA back into the bacterium (*E. coli*) and to have it stably maintained as the cell grows and replicates. It is possible to do this by incorporating the expression system (gene and regulatory sequences) into a plasmid and using the recombinant plasmid to transform bacteria.

3.4 Plasmids

Plasmids are small circular DNA molecules that exist independently of the chromosome in bacteria. Originally discovered through their association with acquired antibiotic resistance, they are exploited as carriers or 'vectors' for foreign DNA sequences in genetic engineering. For use in genetic engineering, plasmids need to include three important DNA sequences, namely an origin of replication which allows the plasmid to replicate and be maintained at levels of, say, 40–200 copies per cell, restriction endonuclease cleavage sites for insertion of the foreign DNA and an antibiotic resistance gene to allow selective growth of plasmid-carrying bacteria in the presence of the antibiotic. For example, the DNA of interest can be inserted into a plasmid which carries the gene encoding β-lactamase, an enzyme which hydrolyses and inactivates the penicillin derivative, ampicillin. Ampicillin prevents bacterial cell growth by inhibiting cell wall synthesis. The expression of β-lactamase therefore relieves this inhibition allowing cell growth and division.

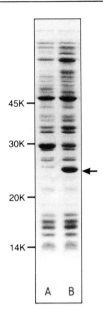

Figure 3.3. A stained SDS-polyacrylamide gel illustrating the expression of a recombinant plasmid-encoded myoglobin fusion protein under the regulation of the λ P_L promoter in an *E. coli* strain expressing the temperature-sensitive λ repressor cI_{857}. Lane A shows an extract from cells grown at 30°C, lane B shows an extract from cells grown at 42°C (3). The arrow indicates the band associated with the recombinant myoglobin and the numbers on the left indicate the positions of migration of molecular mass (in kilodaltons) markers.

3.5 Bacterial transformation

Bacteria that are capable of taking up naked DNA are said to be 'competent' for transformation. Naturally, bacterial cells take up DNA only at low efficiency. This efficiency can be greatly increased by incubating the cells at 0°C in the presence of $CaCl_2$ prior to their exposure to the DNA. Even under these conditions, however, the vast majority of cells will not take up DNA. It is possible to select for those cells that have done so by plating the cells on to ampicillin-containing media. Only those individual cells that have taken up the plasmid will be 'transformed' to antibiotic resistance and be capable of colony formation. These plasmid-bearing cells will of course also carry the gene of interest.

These transformed cells can be grown up and induced to express the gene. Under favourable conditions it is possible to express foreign genes so that the foreign protein represents 5–10% of the total soluble protein of the cell. The protein can then be easily identified in cell lysates as an intense staining band on SDS-polyacrylamide gels (*Figure 3.3*).

4. Site-directed mutagenesis

4.1 Oligonucleotide-directed mutagenesis

A long-standing objective of molecular biologists was to develop a method for making any desired mutation in a sequence of DNA. This was realized in the late 1970s through the pioneering studies of Hutchison and Smith (4), which led to the development of the technique of oligonucleotide-directed mutagenesis. The method, which is conceptually similar to the enzymatic DNA sequencing procedure developed by Sanger (5), is both simple and elegant. The requirements for mutagenesis are single-stranded DNA containing the cloned fragment (of known sequence) that is to be mutated and a synthetic oligonucleotide usually 15–30 nucleotides in length.

Although DNA isolated from most natural sources is double-stranded there are a number of bacterial viruses, for example M13, whose DNA is carried in the single-stranded form. Genetic engineers have been able to take the appropriate M13 sequences responsible for single-stranded DNA production, and incorporate them into plasmids (6). The resultant 'phagemids', as they are referred to, not only retain the useful properties of plasmids for recombinant DNA manipulation but also allow rapid and simple preparation of single-stranded DNA for mutagenesis and DNA sequencing.

The technique of site-directed mutagenesis in its simplest form is illustrated schematically in *Figure 3.4*. It may be used to make insertions, deletions, and substitution mutations. The single-stranded DNA to be mutated and the mutagenic oligonucleotide are mixed and allowed to anneal (form base-pairs) by heating the mix to 65°C followed by cooling to room temperature. The sequence of the oligonucleotide is chosen so that it will form Watson–Crick base-pairs with sequences flanking the site to be mutated and so that there is a mismatch(es) at this site that will direct the mutation. After the annealing step, the oligonucleotide and the single-stranded DNA are used as primer–template for enzyme catalysed complementary strand synthesis *in vitro*. DNA polymerase catalyses the template-directed incorporation of complementary nucleotides and DNA ligase catalyses the ATP-driven covalent closure of the nicked DNA product. The resultant heteroduplex DNA molecule has the potential to give rise to two types of progeny DNA after it has been replicated. Thus, after competent cells have been exposed to the heteroduplex DNA and transformants have been selected following plating, the colonies obtained will carry either mutant or wild-type DNA (*Figure 3.4*). The proportions of mutant and wild-type colonies will depend on the repair and replication processes that take place in the host cell following uptake of the DNA, but usually less than 10% carry the mutated DNA. For this reason it is necessary to screen the colonies for those that carry the mutant sequence.

4.2 Screening for mutants

One of the most powerful possibilities afforded by genetic engineering is to screen

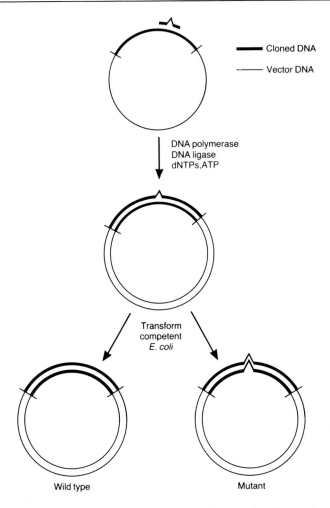

Figure 3.4. Schematic outline of the technique of oligonucleotide-directed mutagenesis.

libraries of DNA fragments for the occurrence of particular sequences of DNA using a radiolabelled probe in a hybridization assay. This technique can also be used to identify mutant DNA sequences from a background of wild-type sequences differing by only a single base-pair. Here the probe that is used is the mutagenic oligonucleotide itself. It will be able to form a perfect Watson–Crick duplex only with the mutated DNA sequences; with the wild-type sequence there will be mismatches.

Transformed cells are applied gridwise to a nitrocellulose filter which has been laid on to an agar plate, and grown overnight to produce colonies. The cells are lysed on the filter and the released DNA is immobilized on the filter by baking in a vacuum oven. The filter is then incubated in the presence of the radiolabelled mutagenic primer, allowing the plasmid DNA sequences to hybridize with the oligonucleotide. Excess oligonucleotide is removed and the filter is washed successively at a series of increasing temperatures. At each step, label bound to DNA on the filter can be detected by autoradiography (*Figure 3.5*). A temperature will be reached at which only the oligonucleotide–mutant DNA hybrid, which is perfectly base-paired, is stable. DNA sequencing of the putative mutant plasmids will confirm the introduction of the mutation.

4.3 Increasing the yield of mutants

Improvements have been made to this simple mutagenesis technique that augment the yield of mutants. One such method takes advantage of an *E. coli* strain that is deficient in two enzymes, deoxyuridine triphosphatase and uracil glycosylase (7). The combined effect of these two deficiencies is that the cells frequently incorporate uracil bases instead of thymine bases into their DNA and are unable to repair the mistake. Template DNA for mutagenesis prepared from these cells will contain on average four to six uracil bases per thousand base-pairs in the molecule. During complementary strand synthesis *in vitro* primed by the mutagenic oligonucleotide, thymine bases are, of course, incorporated correctly into the nascent strand, producing a heteroduplex product in which only one strand (the wild-type) contains uracil bases. When the heteroduplex is reintroduced into a normal bacterial strain that possesses a functional uracil glycosylase, the presence of the uracil bases causes the template strand to be preferentially degraded leaving the *in vitro* synthesized strand containing the mutation intact. It is found that far higher yields of mutants (50–100%) may be obtained with this procedure, obviating the need for colony-blot screening and justifying direct DNA sequencing as a method for identifying mutants.

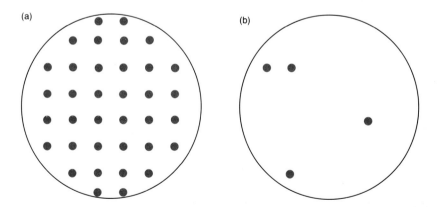

Figure 3.5. Representation of an autoradiograph following colony-blot hybridization and washing of the filter, (a) wash below the melting temperature where all hybrids are stable, and (b) wash at the melting temperature of the primer–mutant DNA hybrid where primer–wild-type DNA hybrids are unstable.

5. Analysis of mutant proteins

5.1 Overview

The analysis of mutant proteins is the most exciting part of protein engineering. The analysis is always comparative: how does the mutant protein differ from the wild-type protein or has the protein acquired the properties it was hoped to engineer into it? For accurate quantitative analysis, assays of the purified mutant proteins should be made in parallel with assays of the purified wild-type protein, so as to eliminate artefacts that may be introduced from using different solutions on different days.

Three outcomes can be anticipated from a mutagenesis experiment. The first is that the activity of the protein is unaffected by the mutation. The interpretation is clear: the residue has no role in the protein's activity. The second is that the modification abolishes activity completely. The third possibility is that the activity is changed. For the latter two possibilities, it is important to establish whether the observed effects are due to an intrinsic loss of activity as opposed to the protein being partly or completely inactive because it is unstable or incorrectly folded.

The nature and details of the assay depend of course on the nature of the protein and the particular property that is being investigated (see Chapters 4 and 5). Keen mechanistic insights have been gained through careful quantitative analysis of the binding of ligands to mutant proteins and of the steady-state kinetics of mutant enzymes. General aspects of these analyses relevant to the discussions in Chapters 4 and 5 are considered here.

5.2 Protein–ligand interactions

The binding of a ligand (L) to a protein (P), to form a complex (PL), may be described as a simple reversible process

$$P + L \underset{k_{-1}}{\overset{k_1}{\rightleftharpoons}} PL \qquad (3.1)$$

involving forward (k_1) and reverse (k_{-1}) rate constants. Here the ligand is not confined to a small metabolite, it may equally well represent a second protein or a DNA molecule. The equilibrium dissociation constant K_S is described by

$$K_S = [P][L]/[PL] = k_{-1}/k_1 = 1/K_A \qquad (3.2)$$

where K_A is the equilibrium association constant. K_S may be determined by measuring both k_1 and k_{-1}, for example by following association and dissociation processes, using a spectroscopic technique. Alternatively, it may be possible to determine K_S by mixing known quantities of protein and ligand and measuring the equilibrium concentrations of the species directly, using an equilibrium dialysis or filter-binding technique. For mutant and wild-type proteins, the ratio of the dissociation constants $K_S(\text{mut})/K_S(\text{wt})$ provides a comparative description of the consequence of the mutation on ligand binding.

Physical biochemists prefer to consider binding processes in terms of the interaction energies involved:

$$\Delta G_B = -RT\ln K_A = RT\ln K_S \quad (3.3)$$

where ΔG_B is the free energy of binding in kcal/mol; (1 calorie = 4.18 Joules), T is the temperature in Kelvin (K), and R is the gas constant (1.99 cal K^{-1} mol^{-1}). Mutant and wild-type proteins can then be considered from the difference in the free energy of binding ($\Delta\Delta G_B$)

$$\Delta\Delta G_B = \Delta G_B(\text{wt}) - \Delta G_B(\text{mut}) \quad (3.4)$$

$$\Delta\Delta G_B = RT\ln K_S(\text{wt}) - RT\ln K_S(\text{mut}) = RT\ln[K_S(\text{wt})/K_S(\text{mut})] \quad (3.5)$$

where $\Delta\Delta G_B$ represents the change in protein–ligand binding energy caused by the mutation. Thus if the mutant protein binds the ligand less tightly than the wild-type protein, $\Delta\Delta G_B$ is algebraically negative.

5.3 Enzyme kinetics

A characteristic of enzyme catalysis is that the enzyme-substrate binding energy is used to enhance the rate of the reaction and to impose substrate specificity. A simple mechanism of enzyme action is shown in Equation 3.6:

$$E + S \underset{k_{-1}}{\overset{k_1}{\rightleftharpoons}} ES \xrightarrow{k_2} E + P \quad (3.6)$$

where E, S, and P represent enzyme, substrate, and product respectively and k_1, k_{-1}, and k_2 are rate constants associated with the indicated steps.

Under conditions where the concentration of enzyme is negligible compared to the concentration of substrate, it is possible to apply a steady-state condition which assumes that the concentration of ES is constant with respect to time. The initial rate of the reaction (V) then depends on the substrate concentration [S], according to the Michaelis–Menten equation,

$$V = k_{\text{cat}}[E_0][S]/(K_M + [S]) = V_{\max}[S]/(K_M + [S]) \quad (3.7)$$

where $k_{\text{cat}} = k_2$, $K_M = (k_{-1} + k_2)/k_1$, $[E_0]$ is the total concentration of enzyme and $V_{\max}$ is the maximum initial reaction velocity. Equation 3.7 applies for more complicated mechanisms than that shown in Equation 3.6, including reactions involving more than one substrate; the constants k_{cat} and K_M then comprise correspondingly more complex sets of rate constants.

A simple way of comparing mutant and wild-type enzymes is to examine the dependence of the rates of the two enzyme catalysed reactions on the substrate concentration and determine the steady-state kinetic parameters k_{cat} and K_M from appropriate graphical plots of initial reaction velocity against substrate

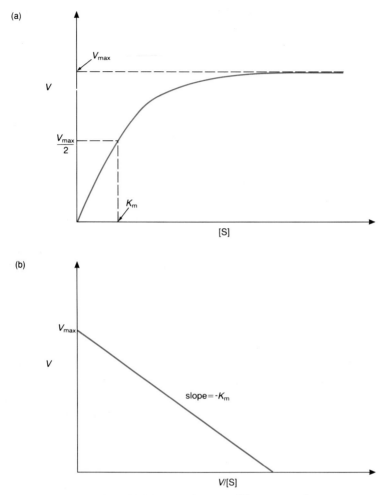

Figure 3.6. (a) Plot of initial reaction rate (V) versus substrate concentration for an enzyme catalysed reaction. (b) Plot of V versus $V/[S]$ for the reaction shown in (a). The gradient represents $-K_M$ and the intercept on the y-axis represents V_{max}.

concentration (*Figure 3.6a* and *b*). This allows the relative rates of the two catalysed reactions to be calculated at any concentration of substrate.

K_M may be defined in a simple way. It is the substrate concentration at which the initial reaction velocity is half the maximum initial reaction velocity (*Figure 3.6a*) and can be determined without a prior knowledge of the total enzyme concentration. If, in Equation 3.6, $k_{-1} \gg k_2$, which in practice is quite common, then K_M approximates to the equilibrium dissociation constant ($K_S = k_{-1}/k_1$) for the enzyme–substrate complex. For many more complex mechanisms, including those discussed in Chapter 4, K_M often represents an apparent dissociation constant for all of the enzyme-bound species as in the equation:

$$K_M = [E_f][S]/\Sigma[ES] \qquad (3.8)$$

where $[E_f]$ is the concentration of the free enzyme species and $\Sigma[ES]$ is the sum of the concentrations of all the enzyme-bound species (8).

The constant k_{cat} represents the number of molecules converted to product per unit time by each active site at saturating substrate concentrations. It is calculated from the quotient of the maximum initial reaction velocity (at infinite substrate concentration) and the enzyme concentration ($k_{cat} = V_{max}/[E_0]$). The measurement of k_{cat} is therefore sensitive to the accuracy of the method by which the concentration of active sites is determined. For many enzymes, the concentration is inferred from measurements of the total protein concentration in a purified preparation. Such measurements take no account of the proportion of the enzyme which is actually *active* and can therefore be subject to significant inaccuracies, because of the intrinsic instability of many proteins with respect to denaturation. This problem is overcome if an active-site titration assay is available. Accurate active-site titrations can be performed if substrates (e.g. for tyrosyl-tRNA synthetase) or inhibitors (e.g. for serine proteases) bind stably and stoichiometrically at the active site. Quantitative comparisons of k_{cat} can then be made permitting detailed interpretation of the consequences of the structural changes.

5.4 The energetics of enzyme catalysis

The energetic contributions that functional groups on enzymes make to catalysis can be evaluated by examining the steady-state kinetic parameters of appropriate mutant enzymes in more detail. The Michaelis–Menten equation may be recast by substituting Equation 3.8 into Equation 3.7 together with the knowledge that $[E_0] = \Sigma[ES] + [E_f]$ to give Equation 3.9 (8)

$$V = (k_{cat}/K_M)[E_f][S] \qquad (3.9)$$

This is a useful equation as k_{cat}/K_M can be seen to represent a second-order rate constant. It may therefore be used to describe the process of proceeding from the reactants to the transition state of the reaction, $ES^{\neq}$, Equation 3.10.

$$E + S \underset{\Delta G_T^{\neq}}{\overset{k_{cat}/K_M}{\longrightarrow}} ES^{\neq} \qquad (3.10)$$

The transition state is the highest energy state on the reaction co-ordinate and is denoted by the superscript $\neq$. The free energy barrier associated with this process, $\Delta G_T^{\neq}$, may therefore be calculated using the Equation 3.11

$$\Delta G_T^{\neq} = RT\ln(k_B T/h) - RT\ln(k_{cat}/K_M) \qquad (3.11)$$

where k_B is the Boltzmann constant and h is Planck's constant.

When comparing mutant and wild-type enzymes, the difference in $\Delta G_T^{\neq}$ for the wild-type and mutant proteins, or the change in the enzyme–substrate binding energy in the transition state is given by

$$\Delta\Delta G_T^{\ddagger} = \Delta G_T^{\ddagger}(\text{wt}) - \Delta G_T^{\ddagger}(\text{mut}) \quad (3.12)$$

Combining Equations 3.12 and 3.11 gives

$$\Delta\Delta G_T^{\ddagger} = RT\ln\{(k_{cat}/K_M)_{mut}/(k_{cat}/K_M)_{wt}\} \quad (3.13)$$

where mut and wt denote mutant and wild-type enzymes, respectively (9). $\Delta\Delta G_T^{\ddagger}$ thus represents the change in the enzyme-transition state binding energy caused by the mutation. On the basis of Equation 3.13, $\Delta\Delta G_T^{\ddagger}$ is algebraically negative for mutations with destabilizing effects on transition state binding, that is, mutations that lower the value of k_{cat}/K_M.

Penetrating insights can be made if $\Delta\Delta G_T^{\ddagger}$ can be interpreted in terms of the structural consequences of the mutation, that is to attribute the energy difference to a specific chemical interaction between the enzyme and the substrate in its transition state. In a typical site-directed mutagenesis experiment, the interaction that a side-chain group R on the enzyme makes with the substrate is investigated by altering the amino acid residue to remove the R group. It is then tempting to ascribe $\Delta\Delta G_T^{\ddagger}$ of Equation 3.13, to the interaction that R makes with the substrate in the transition state. Although this may often be true, such interpretations must be made with caution because the structural perturbations in the mutant enzyme may not be confined to the site of mutation but may be propagated through the structure so that other enzyme–substrate interactions are affected. This can be tested in a number of ways, including through further site-directed mutagenesis experiments or X-ray analysis of the mutant enzyme.

It is possible to dissect the energetics of the reaction further by considering the effects of the mutation on enzyme–substrate binding in the ground state as well as in the transition state as shown in *Figure 3.7*. $\Delta G_T^{\ddagger}$ is composed of an energetically favourable term ΔG_B associated with enzyme–substrate binding and an energetically unfavourable term $\Delta G^{\neq}$ associated with the chemical activation step (8) (*Figure 3.7*).

$$\Delta G_T^{\ddagger} = \Delta G_B + \Delta G^{\neq} \quad (3.14)$$

A mutation in the enzyme may affect either or both of these free energy differences by changing the stability of the free enzyme, the enzyme–substrate complex in its ground state or the enzyme–substrate complex in its transition state. For enzymes whose kinetic behaviour satisfies Equation 3.8, the constants k_{cat} and K_M of the Michaelis-Menton equation provide information on the binding and catalytic steps of the reaction, and K_M can be used to calculate ΔG_B and k_{cat} to calculate $\Delta G^{\neq}$. If the interaction of R with the substrate is made uniformly in the ground state and the transition state, removal of this group (*Figure 3.7*, orange line) should increase K_M without affecting k_{cat}; ΔG_B will be

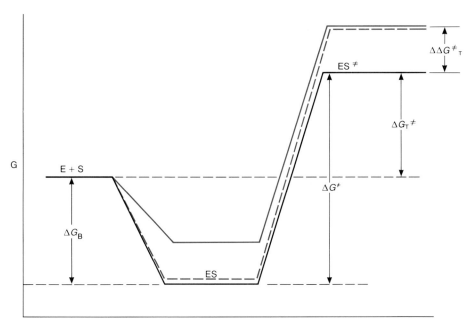

Figure 3.7. Free energy profile of the course of an enzyme catalysed reaction. The orange line shows the effect of a mutation which causes a uniform decrease in the enzyme–substrate binding energy while the dashed line shows the effect for a mutant that lowers the enzyme–substrate binding energy only in the transition state.

greater (less negative) and $\Delta G^{\neq}$ will be unaffected. In contrast, if R binds the substrate only in its transition state, its removal should not affect K_M but lower k_{cat}; ΔG_B will be unaffected, but $\Delta G^{\neq}$ will be increased (*Figure 3.7*, dashed line).

A fuller general description of the effects of the mutation requires the measurement of each of the microscopic rate constants (the small ks in Equation 3.6) any of which may be altered as a consequence of the structural change. This can be done by following the pre-steady-state kinetic behaviour but this is outside the scope of the discussion here. In combining these rate constants together as just two observables in the steady-state analysis, the subtle effects of the mutation are often not apparent.

6. Further reading

Zoller,M.J. and Smith,M. (1983) *Methods Enzymol.*, **100**, 468.
Rossi,J. and Zoller,M. (1987) In Oxender,D.L. and Fox,C.F. (ed.), *Protein Engineering*. Alan R. Liss, New York.
Ptashne,M. (1986) *A Genetic Switch; Gene Control in Phage* λ. Blackwell Scientific Publications and Cell Press, Cambridge, MA, USA.
Cornish-Bowden,A. and Wharton,C.W. (1988) In Rickwood,D. (ed.), *Enzyme Kinetics: In Focus*. IRL Press, Oxford and Washington.
Fersht,A.R. (1985) *Enyzme Structure and Mechanism*. Freeman, New York.

7. References

1. Beebee,T. and Burke,J. (1988) In Rickwood,D (ed.), *Gene Structure and Expression: In Focus*. IRL Press, Oxford and Washington.
2. Williams,J.G. and Patient,R.K. (1988) In Rickwood,D. (ed.), *Genetic Engineering: In Focus*. IRL Press, Oxford and Washington.
3. Dodson,G.G., Hubbard,R.E., Oldfield,T.J., Smerdon,S.J. and Wilkinson,A.J. (1988) *Protein Engineering,* **2**, 233.
4. Hutchison,C.A., Phillips,S.A., Edgell,M.H., Gillam,S., Jahnke,P. and Smith,M. (1978) *J. Biol. Chem.,* **253**, 6551.
5. Sanger,F., Nicklen,S. and Coulson,A.R. (1977) *Proc. Natl. Acad. Sci. USA,* **74**, 5463.
6. Dente,L., Cesarini,G. and Cortese,R. (1983) *Nucleic Acids Res.,* **11**, 1645.
7. Kunkel,T. (1985) *Proc. Natl. Acad. Sci. USA,* **82**, 488.
8. Fersht,A.R. (1974) *Proc. R. Soc. Lond. B,* **187**, 397.
9. Wilkinson,A.J., Fersht,A.R., Blow,D.M. and Winter,G. (1983) *Biochemistry,* **22**, 3581.

4

Protein engineering: site-directed mutagenesis as a probe of function

1. Introduction

In this chapter the ways in which the techniques of genetic engineering may be employed for the study of protein function are described prior to considering their application to problems with medical and/or industrial implications in the following chapter.

A great many techniques have been applied to the study of how enzymes function. Significant among these have been studies with chemically synthesized substrate analogues which have been used to define groups and surfaces on substrates that are important for interaction with the enzyme. The reciprocal approach of specifically modifying a group on the enzyme by chemical means is generally not possible, due to the large number of such groups in the overall structure. Where it has proved possible to make such specific modifications, the results require cautious interpretation because, as well as blocking a potentially reactive group, chemical modification almost always increases local steric bulk.

The peptidase action of carboxypeptidase-A is sensitive to nitration of the aromatic ring of tyrosine-248. This observation was used to reinforce the idea that the phenolic hydroxyl group of tyrosine-248 acts as an acid in catalysis, contributing a proton to the incipient amine formed during cleavage of the peptide bond. However, because substitution of this tyrosine for phenylalanine has no effect on the enzyme activity, this mechanism can now be discounted (1). Clearly, with the advent of the technique of site-directed mutagenesis and the possibility of introducing any desired amino acid substitution into a protein whose gene has been cloned and sequenced, there is the opportunity to carry out rigorous tests on proposed mechanisms.

2. Tyrosyl-tRNA synthetase

2.1 Reaction and properties

Tyrosyl-tRNA synthetase from *Bacillus stearothermophilus* was among the first enzymes to be probed using the technique of site-directed mutagenesis (2), and a great deal has been learned from these studies. As will be illustrated in the following sections, the role of hydrogen-bonding in binding, specificity, and catalysis has been explored in considerable detail. The enzyme, a dimer of identical subunits ($M_r = 2 \times 47\,500$) catalyses the two-step aminoacylation of tRNA^{Tyr} with tyrosine (Equations 4.1 and 4.2), an essential first step in protein synthesis.

$$E + Tyr + ATP \rightleftharpoons [E \cdot Tyr\text{-}AMP] + PP_i \qquad (4.1)$$

$$[E \cdot Tyr\text{-}AMP] + tRNA^{Tyr} \longrightarrow E + Tyr\text{-}tRNA^{Tyr} + AMP \qquad (4.2)$$

Tyrosine is first activated by reaction with ATP to form the enzyme-bound intermediate, tyrosyl-adenylate, with the liberation of inorganic pyrophosphate. The amino acid is then transferred to the 3′-hydroxyl group of the 3′ adenosine of the tRNA with release of AMP. The progress of both the overall tRNA charging reaction and the activation step (Equation 4.1) of the reaction may be followed using filter-binding assays with radiolabelled substrates.

The enzyme-bound intermediate is unusually stable. When the enzyme is incubated with tyrosine and ATP in the presence of inorganic pyrophosphatase, which hydrolyses the pyrophosphate formed in the reaction shown in Equation 4.1 and therefore blocks the reverse reaction, the enzyme–tyrosyl-adenylate complex is formed stoichiometrically. This has two important consequences. First, it allows the concentration of active sites in enzyme preparations to be measured accurately using a simple filter-binding assay which uses radiolabelled tyrosine. For a mutant enzyme it is important to establish that changes in V_{max} are caused by changes in k_{cat} and not for the trivial reason that there is a change in the proportion of the enzyme in the mutant preparation which remains functional. Second, it makes it possible to prepare and collect X-ray diffraction data on crystals of the enzyme–intermediate complex, allowing the determination of the structure of this complex and the identification of residues on the enzyme which are in contact with the intermediate (3). *Figure 4.1* shows that there are 11 hydrogen-bonding contacts made between the enzyme and the intermediate of which eight are made by side-chains and amenable to substitution using site-directed mutagenesis.

The gene encoding the *B. stearothermophilus* tyrosyl-tRNA synthetase has been cloned, sequenced, and incorporated into the bacteriophage M13 for expression and mutagenesis. Overproduction of the enzyme is achieved following infection of *E. coli* with the recombinant bacteriophage. High expression is directed by the natural promoter. Multiple copies of the gene in each cell presumably account for the excellent yields of recombinant protein obtained.

2.2 Probing the role of cysteine-35 in enzyme–substrate binding

Hydrogen bonds are interactions that involve the sharing of a hydrogen atom by two electronegative atoms such as $>\!N\!-\!H\cdots O\!=\!C\!<$. The atom to which

Site-directed mutagenesis as a probe of function   37

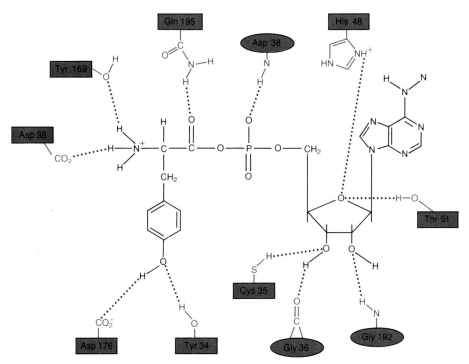

Figure 4.1. Schematic diagram of the enzyme-bound intermediate, illustrating residues on the enzyme (orange) which may form hydrogen-bonding interactions with the tyrosyl-adenylate (black) (3, 4). Side-chain contacts are denoted by rectangular boxes and main-chain contacts by rounded boxes.

the hydrogen is formally bonded is described as the hydrogen-bond donor, here >N—H, while the second electronegative atom, here the oxygen of the >C=O, whose distance from the hydrogen atom is usually less than the van der Waals contact distance, is referred to as the hydrogen-bond acceptor. Hydrogen-bonding interactions are enormously important in biology. They play crucial roles in determining the structures of proteins and DNA, and in catalysing chemical interconversions by solvating charges formed in the transition states of reactions. They are weak enough to form reversibly but strong enough to impose specificity.

In the first mutagenesis studies on tyrosyl-tRNA synthetase (TyrRS), attention focused on residues that interact with the ribose sugar of the ATP and in particular on the hydrogen-bonding interaction cysteine-35 appears to make with the 3'-OH group of the sugar in the crystal structure. Cysteine-35 is remote from the centre of the reaction; the important chemistry occurs at the phosphorous centre (*Figure 4.1*). This residue was modified to serine and glycine respectively, that is, first by changing the nature of the hydrogen bonding group, and second by removing it altogether (2,5). These mutations were intended to alter the enzyme structure minimally so that activity is not destroyed. This allows quantitative kinetic measurements to be made on the mutant enzymes to evaluate the role of the hydrogen bond in catalysis. Steady-state kinetic measurements

Table 4.1. Effects of mutation of cysteine-35 of tyrosyl-tRNA synthetase on the steady-state kinetic parameters and on the rate of reaction at 298K (2).

Enzyme	k_{cat} (s^{-1})	K_M (mM)	k_{cat}/K_M ($s^{-1}\,M^{-1}$)	Rate of reaction* ($\mu M s^{-1}$) 0.1 mM	1 mM	10 mM	$\Delta\Delta G_T^{\ddagger}$ (kcal/mol)
TyrRS(Cys 35)	7.6	0.9	8400	0.76	4.0	7.0	–
TyrRS(Ser 35)	2.4	2.4	1000	0.10	0.71	1.9	–1.3
TyrRS(Gly 35)	2.8	2.6	1120	0.10	0.78	2.2	–1.2

*Rates of reaction calculated at 1 μM enzyme at the ATP concentrations shown.

were made on both wild-type and mutant enzymes and the results are shown in *Table 4.1* for the activation reaction (Equation 4.1). The effects on the overall aminoacylation reaction are both qualitatively and quantitatively similar for the two mutants.

The data in *Table 4.1* show that the two mutant enzymes TyrRS(Ser 35) and TyrRS(Gly 35) are less efficient catalysts than the wild-type enzymes, owing to increases in K_M and decreases in k_{cat}. The extent to which the mutant enzyme catalysed reactions are slower depends on the ATP concentration at which the assay is carried out. According to the Michaelis–Menten equation (Equation 3.7), at saturating ATP concentrations the rates are proportional to k_{cat}, thus the mutants are three- to fourfold slower. At concentrations of ATP below K_M the rates are proportional to k_{cat}/K_M, the mutants are thus eightfold slower. In the physiological environment, where only a narrow range of ATP concentrations of around one millimolar are encountered, the wild-type enzyme is more efficient by a factor of five- to sixfold.

As discussed in Chapter 3, the quantity of k_{cat}/K_M may be used to calculate the change in the free energy of the enzyme and the substrate as they proceed to the transition state of the reaction. For two enzymes, mutant and wild-type, that differ by the removal of a group R, it is possible to calculate the contribution of R to the binding and stabilization of the transition state, using Equation 3.13. It is found that the glycine and serine substitutions decrease the binding energy in the transition state complex by about 1.2 kcal/mol (*Table 4.1*).

Is it possible to interpret these results in terms of the structures involved? For TyrRS(Gly 35), it is tempting to attribute the loss of binding energy in the transition state to removal of the hydrogen bonding interaction made between the cysteine —SH group and the ribose hydroxyl group in the substrate (*Figure 4.1*). The observations for TyrRS(Ser 35) are more puzzling. The oxygen atom of an aliphatic hydroxyl group is more electronegative than the sulphur atom of a sulphydryl group, so that the —OH group of serine is able to form stronger hydrogen bonds than the —SH group of cysteine. At residue 35 however, it is observed that the serine enzyme binds the transition state of the substrate more weakly than the cysteine enzyme. This can be understood if the role of the solvent is taken into account, rather than oversimplifying the description of the binding process as in Equation 4.3:

$$\text{Cys} - \text{S}^{\diagup \text{H}} + \overset{\text{Ribose}}{\underset{\text{H}}{|}} \overset{|}{\text{O}} \rightleftharpoons \text{Cys} - \text{S}^{\diagup \text{H} \cdots \text{O}^{\diagdown \text{Ribose}}_{\diagdown \text{H}}} \qquad (4.3)$$

Although serine can form stronger hydrogen bonds than cysteine, the reaction takes place in water, so that active site groups on the enzyme will be solvated in the absence of the substrate. Thus, the binding of substrate (or here the transition state of the substrate) to the enzyme must be seen as an exchange process in which the solvent is displaced as in Equation 4.4 rather than simple

$$E-H \cdots OH_2 + HO-H \cdots B-S \rightleftharpoons E-H \cdots B-S + HO-H \cdots OH_2 \qquad (4.4)$$

substrate binding as in Equation 4.3. In this and subsequent equations, $E-H$ represents a hydrogen-bond donor on the enzyme such as the $-SH$ of cysteine-35 or the $-OH$ group of serine-35 and $B-S$ represents a hydrogen-bond acceptor on the substrate such as the oxygen atom of the 3' hydroxyl group on the ribose. It can be seen that the absolute strength of the hydrogen bond that is made in Equation 4.4 is not important: a stronger hydrogen-bonding group on the enzyme will also form a strong hydrogen bond with the solvent. This strong hydrogen bond with the solvent has to be sacrificed in forming a good hydrogen bond with the substrate.

The fact that TyrRS(Ser 35) binds the transition state more weakly is due to differences in the size and geometry of the $-SH$ and $-OH$ groups (*Figure 2.2*). If the structures of serine and cysteine are superimposed at their α-carbon atoms, the position of the oxygen atom in serine is some 0.5 Å short of the position of the sulphur atom in cysteine. In the absence of structural reorganization in the enzyme the hydrogen bond formed between the serine $-OH$ and the substrate will therefore be 0.5 Å longer; evolution has presumably optimized the geometry for the enzyme-substrate hydrogen bond involving cysteine-35. The lengthening of a hydrogen bond by 0.5 Å is sufficient to weaken the hydrogen bond substantially so that on binding the substrate, TyrRS(Ser 35) will be swapping a good hydrogen bond with the solvent for a poor one with the substrate. Consequently TyrRS (Ser 35) is a poorer enzyme. As the serine and glycine mutant enzymes appear to bind the transition state equally tightly it appears that this poor H-bond contributes nothing to catalysis.

2.3 Hydrogen-bonding and specificity

The above strategy has been extended into a systematic study of residues of tyrosyl-tRNA synthetase that appear from the crystal structure to make hydrogen-bonding interactions with the substrate (*Figure 4.1*). The contributions that hydrogen-bonding groups make to the binding of the transition state of the reaction have been evaluated using Equation 3.13 and are shown in *Table 4.2*

Table 4.2. Relative binding energies of groups in tyrosyl-tRNA synthetase inferred from comparisons between mutant and wild-type enzymes at 298K. The data are taken from ref. 4

Comparison		Substrate	$\Delta\Delta G_T^{\ddagger}$ (kcal/mol)
Enzyme 1	vs Enzyme 2		
Phe 34	Tyr 34	Tyr	0.52
Gly 35	Cys 35	ATP	1.14
Ala 51	Cys 51	ATP	0.47
Gly 48	Asn 48	ATP	0.77
Gly 48	His 48	ATP	0.96
Ser 35	Cys 35	ATP	1.18
Phe 169	Tyr 169	Tyr	3.72
Gly 195	Gln 195	Tyr	4.49
Gly 35	Ser 35	ATP	−0.04
Ala 51	Thr 51	ATP	−0.44

(4). These energies measured in solution fall into two categories. First, deletion of a hydrogen-bond donor or acceptor on the enzyme that interacts with an uncharged donor or acceptor on the substrate weakens binding energy by 0.5–1.5 kcal/mol. This is considerably less than the range of interaction energies calculated for the formation of such hydrogen bonds between molecules in a vacuum (3–6 kcal/mol). Second, the deletion of a group on the enzyme that interacts with a charged group on the substrate causes the loss of about 4.0 kcal/mol, again less than the 15–20 kcal/mol for a similar interaction *in vacuo*.

In terms of the specificity that such binding energies can confer, 0.05–1.5 kcal/mol can give two- to 12-fold discrimination in rate (k_{cat}/K_M), while 4.0 kcal/mol can give 1000-fold discriminations (Equation 3.13). It is possible to understand the chemical basis of this distinction by again considering the role of the solvent in the binding process and counting the number and types of hydrogen bonds formed on each side of the appropriate equation. Equation 4.4 described the situation for the formation of an enzyme–substrate hydrogen bond between an uncharged donor and an uncharged acceptor. Deleting the hydrogen-bonding group on the enzyme, as in Equation 4.5:

$$E\ OH_2 + HO-H\cdots B-S \rightleftharpoons [E\ B-S] + HO-H\cdots OH_2 \qquad (4.5)$$

does not disturb the overall balance of the equation; that is, there is now one hydrogen bond on each side of the equation rather than two. It is found that this type of mutation results in the loss of 0.5–1.5 kcal/mol of binding energy in solution, implying that the presence of the hydrogen bond in the wild type enzyme contributes 0.5–1.5 kcal in binding energy. From the analysis above, it is surprising that formation of such hydrogen bonds between an enzyme and a substrate confers any binding energy at all. The explanation for this requires a detailed knowledge of the precise interactions made by the water molecule and the substrate in the mutant enzyme.

The situation is not very different when a charged side-chain on the enzyme that interacts with a hydrogen-bonding group on the substrate is deleted as in Equation 4.6:

$$E-H^+ \cdots OH_2 + HO-H \cdots B-S \rightleftharpoons E-H^+ \cdots B-S + HO-H \cdots OH_2 \quad (4.6)$$

Here, a charge–dipole interaction is taken away from both sides of the equation to give Equation 4.5 again. Thus, for the deletion of the charged imidazole side-chain of histidine-48 which interacts with the ribose ring oxygen, only 0.5 kcal/mol of binding energy is lost.

A different outcome is apparent for the case of deleting a polar side-chain on the enzyme that forms a hydrogen bond with a charged group on the substrate (Equations 4.7 and 4.8):

$$E-H \cdots OH_2 + HO-H \cdots {}^-B-S \rightleftharpoons E-H \cdots {}^-B-S + HO-H \cdots OH_2 \quad (4.7)$$

$$E \quad OH_2 + HO-H \cdots {}^-B-S \rightleftharpoons [E \quad {}^-B-S] + HO-H \cdots OH_2 \quad (4.8)$$

Here, it can be seen that the charge–dipole interaction on the left-hand side of Equation 4.8 is counterbalanced by only a dipole–dipole interaction on the right-hand side of this equation. This is the situation for removing the polar groups of tyrosine-169 and glutamine-195 (although glutamine-195 does not interact with a charged group in the intermediate complex (*Figure 4.1*), it most likely interacts with the charged carboxylate ion of tyrosine in the transition state of the reaction). In binding the substrate and the transition state of the substrate the mutant enzyme creates unsolvated charge. This is much more costly in free energy, approximately 4 kcal/mol. In summary, hydrogen bonds generally confer small binding energy, except where they solvate charged groups.

Tyrosyl-tRNA synthetase must discriminate efficiently between the cognate amino acid tyrosine and the non-cognate substrate phenylalanine. This is essential for faithful translation of the genetic code in the cell. Another way of saying this is that TyrRS must catalyse the aminoacylation reaction much more rapidly for tyrosine than it does for phenylalanine—therefore the enzyme-substrate binding energy in the transition state must be correspondingly greater for tyrosine than for phenylalanine. Structurally, phenylalanine and tyrosine differ only in the extra phenolic hydroxyl group possessed by tyrosine (*Figure 2.2*); thus steric exclusion is not possible. So how does TyrRS discriminate in favour of tyrosine and against phenylalanine? As might be expected from the preceding arguments, it does so using the principle of avoiding unsolvated charge to discourage the binding of phenylalanine. A charged carboxylate group, that of aspartate-176, resides at the bottom of the tyrosine-binding pocket (*Figure 4.1*). Of the two amino acids, only tyrosine can solvate this charged carboxylate on binding. This interaction alone should favour tyrosine binding by a factor of 1000 or more.

To confirm this thinking, a most interesting mutant to make would be one where aspartate-176 is replaced with asparagine. It is predicted that TyrRS(Asn 176) would discriminate very poorly between tyrosine and phenylalanine. Unfortunately, it is also anticipated that there would be considerable problems expressing such a mutant. Its presence in the cell would presumably result in high frequency of misincorporation of phenylalanine in place of tyrosine in protein synthesis with chaotic consequences.

2.4 Catalysis of tyrosyl-adenylate formation

Substitution of the side-chains of residues implicated in the crystal structure as interacting with the intermediate complex does not reveal how the enzyme achieves its rate enhancement for tyrosyl-adenylate formation. As noted earlier, molecular modelling can be a penetrating extension of protein crystallography, allowing predictions to be made that can be tested by site-directed mutagenesis. A good example of this approach is the elucidation of the catalytic roles of threonine-40 and histidine-45 in the formation of tyrosyl-adenylate from tyrosine and ATP (6).

The crystal structure of the enzyme-intermediate complex provides no information as to how the substrate ATP is bound by the enzyme prior to the nucleophilic attack by the tyrosine carboxylate group; in particular because the β- and γ-phosphates of the ATP are not present in this complex. Stereochemical studies of the activation reaction show that the nucleophilic attack proceeds with inversion of configuration at the α-phosphorous atom implying that there is a penta-coordinate transition state (7). This transition state was modelled by introducing the penta-coordinate geometry at the α-phosphorous atom and then building the β- and γ-phosphate groups on to the crystallographic model using computer graphics (*Figure 4.2*). The possible conformations of these phosphate groups were then explored through rotations about the P – O bonds, and groups on the enzyme were identified that could interact with the phosphates to stabilize the transition state. The side-chains of two residues threonine-40 and histidine-45 appear to be well-placed to make stabilizing charge – dipole interactions through their β-OH and ϵ-NH groups respectively with the anionic γ-phosphate group (*Figure 4.2*).

The involvement of these two residues was tested through the construction of two mutants TyrRS(Ala 40) and TyrRS(Gly 45). The rates of formation of tyrosyl-adenylate were decreased dramatically for these enzymes (by 7000-fold and 200-fold respectively). For the double mutant TyrRS(Ala 40, Gly 45) the rate was decreased by a factor of 300 000.

These studies, together with studies of the reaction of the model compound shown in *Figure 4.3*, show how tyrosyl-tRNA synthetase achieves its enormous enhancement of the rate of formation of tyrosyl-adenylate (6, 8). There are two major and separable factors involved. The first is that the enzyme uses the binding energy of the two substrates to pay for the loss in entropy that results when two molecules bind together. It is well-established in physical-organic chemistry that the rates of unimolecular reactions are considerably faster (up to 10^8-fold) than their bimolecular equivalents. The cyclization of the compound in *Figure 4.3* involves nucleophilic attack of an ionized carboxyl group on a

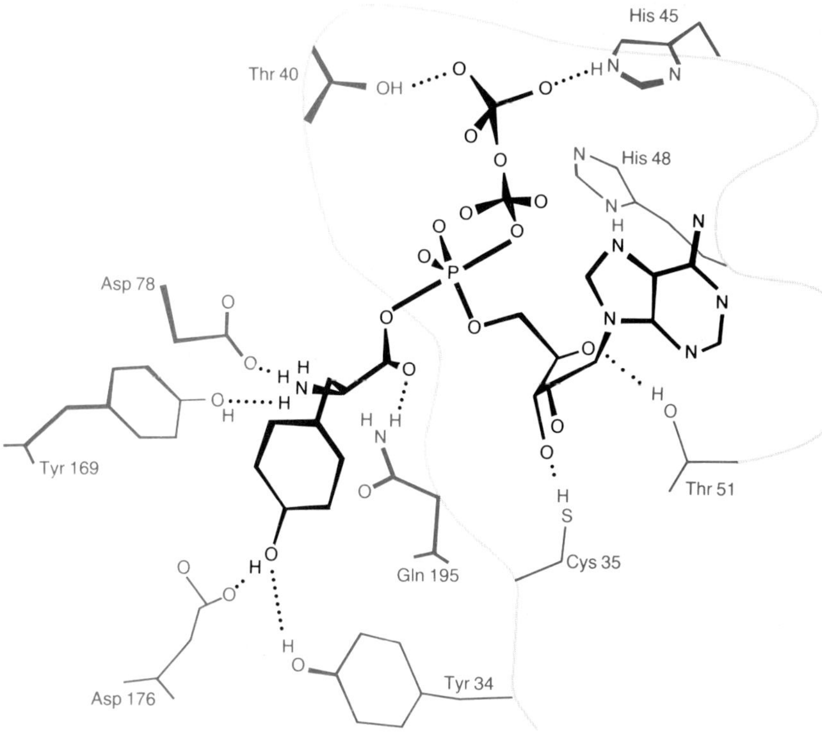

Figure 4.2. Model-building of the penta-coordinate transition state (black) in tyrosyl-adenylate formation into the crystal structure of TyrRS (orange) (After 6).

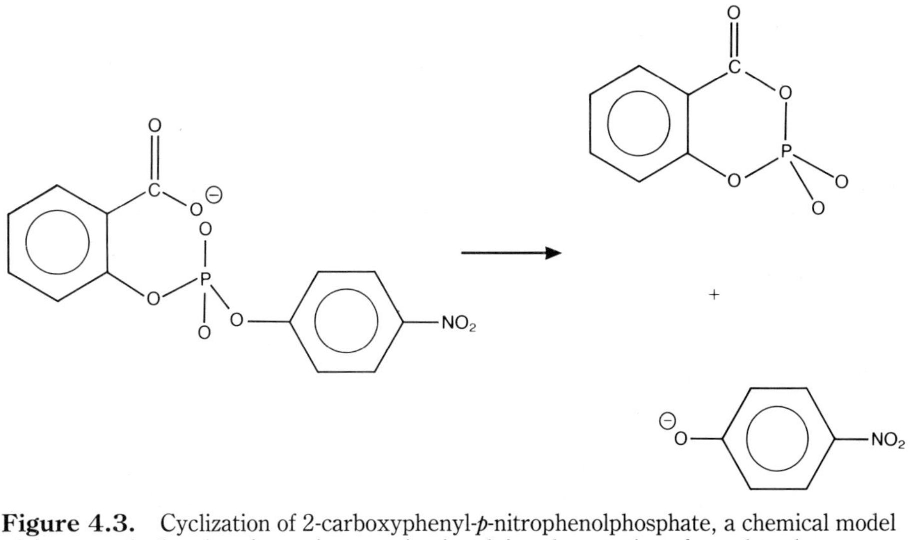

Figure 4.3. Cyclization of 2-carboxyphenyl-*p*-nitrophenolphosphate, a chemical model of the tyrosyl-adenylate formation reaction involving the reaction of a carboxylate group with an adjacent phosphate ester (6, 8).

phosphate ester that possesses a good leaving group. It provides an excellent chemical model for the formation of tyrosyl-adenylate catalysed by TyrRS. It is estimated that the rate of this unimolecular reaction is similar to the rate of tyrosyl-adenylate formation catalysed by the TyrRS(Ala 40, Gly 45) double mutant. The remaining 10^5-fold rate enhancement of the wild-type enzyme arises from the relief of strain in the transition state through the realization of new interactions of the γ-phosphate group of the ATP with the enzyme in the transition state.

In summary, the enzyme combines with tyrosine and ATP. It uses the binding energy of the substrate to compensate for the entropy loss of two molecules, tyrosine and ATP, condensing to form one. In the initial complex the phosphates are unstrained and interact weakly or not at all with threonine-40 and histidine-45. The kinetic measurements indicate that the binding of ATP in its ground state is not affected by mutation at either position 40 or 45. Nucleophilic attack of the carboxylate group of tyrosine causes a stereochemical rearrangement at the α-phosphorous atom. The effects of this configurational reorganization are propagated through the P–O bonds resulting in a rotation of the γ-phosphate group such that it makes strong charge–dipole interactions with threonine-40 and histidine-45. The unfavourable free energy change associated with the formation of a penta-covalent phosphorous is thus partially offset by the realization of the favourable free energy associated with the newly formed strong enzyme–substrate interactions.

3. The serine proteases

3.1 Introduction

The serine proteases are a family of enzymes that catalyse the cleavage of peptide bonds in proteins, an important and recurring process in biology. The common feature of these enzymes to which they owe their name is the presence of a uniquely reactive serine residue that is modified irreversibly by di-isopropyl fluorophosphate (DFP) inactivating the enzyme. Studies relating the structure of serine proteases to their function form the core of enzymology and have done much to influence current ideas on how enzymes work.

Serine proteases enhance the rate of peptide-bond hydrolysis (Equation 4.9) by factors of $10^9 - 10^{10}$ relative to the uncatalysed reactions.

$$P_N-CHR-CO-NH-P_C + H_2O \longrightarrow P_N-CHR-CO_2H + P_C-NH_2 \qquad (4.9)$$

They will also catalyse the hydrolysis of synthetic peptide paranitroanilides, providing a convenient and simple assay of enzyme activity. The formation of the nitroaniline product, which absorbs strongly at 410 nm, may be followed spectrophotometrically.

The characteristic cleavage specificity of a particular serine protease is determined by the side-chain (R in Equation 4.9) of the residue on the N-terminal side of the scissile peptide bond. The pancreatic protease chymotrypsin cleaves after the aromatic residues phenylalanine, tyrosine, or tryptophan while trypsin cleaves after lysine or arginine residues. The two extracellular bacterial proteases, subtilisin and α-lytic protease, discussed later, have broader specificities but prefer hydrophobic R groups.

3.2 Mechanism of action

While the specificity of the cleavage reactions varies, the underlying chemistry of peptide-bond hydrolysis remains the same. Kinetic studies and studies with synthetic substrate analogues and inhibitors have demonstrated that the reaction proceeds via an acyl-enzyme intermediate according to the scheme below:

$$E + S \rightleftharpoons [ES] \longrightarrow P_C + [ES'] \xrightarrow{H_2O} E + P_N \qquad (4.10)$$

The enzyme E and the substrate S combine to form a Michaelis complex ES. A nucleophilic group on the enzyme, the hydroxyl group of the reactive serine, then attacks the carbonyl carbon of the scissile peptide bond to produce the acyl-enzyme intermediate ES', with release of the C-terminal peptide, P_C. The N-terminal peptide, P_N, is released by the reverse reaction in which water replaces the enzyme as the nucleophile and the N-terminal peptide acts as the leaving group. This mechanism is outlined in *Figure 4.4*.

The crystal structures of serine proteases and their complexes with inhibitors reveal the stereochemical arrangement of the catalytic residues and the three-dimensional nature of the substrate-binding pocket; this gives an insight into their mechanism and specificity (9). The anatomy of the active-site region of chymotrypsin is shown in *Figure 4.5*. Serine-195 is in close proximity to the

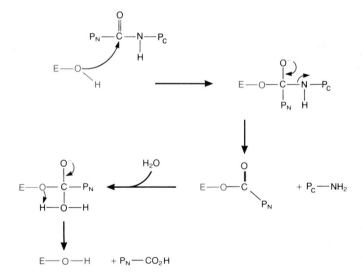

Figure 4.4. Mechanism of serine protease catalysis showing the formation and hydrolysis of the acyl-enzyme.

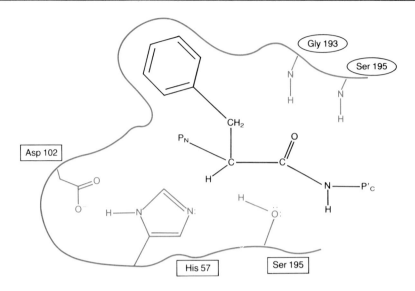

Figure 4.5. Schematic diagram of the active site of chymotrypsin with a peptide substrate bound, illustrating the side-chains (rectangular boxes) of the Ser-His-Asp catalytic triad and the main chain (rounded boxes) contacts in the oxyanion binding site. Note serine-195 makes two contacts with the substrate.

carbonyl carbon of the scissile peptide bond. This was expected because serine-195 had been identified as the residue modified by DFP. In hydrogen-bonding distance of this serine is histidine-57, whose presence at the active site had been inferred from studies with the affinity reagent tosyl-phenylalanyl chloromethyl ketone (TPCK). TPCK chemically modifies histidine-57 with irreversible loss of catalytic activity. What was not predicted, however, was the presence of aspartate-102 closely apposed to histidine-57. These active-site residues are invariant and exactly superimposable in the X-ray structures of other serine proteases, including subtilisin. This is remarkable because subtilisin shows absolutely no amino acid sequence homology with chymotrypsin. Indeed, the corresponding residues in subtilisin, aspartate-32, histidine-64, and serine-221 appear in a different order in the polypeptide sequence. The Ser-His-Asp arrangement in *Figure 4.5* is referred to as the catalytic triad. The pH/activity profile of chymotrypsin indicates that the enzyme is inactivated with the protonation of a group with a pK_a of 7. This group has been shown by neutron diffraction studies and by NMR to be histidine-57.

A 'charge relay' mechanism (9), illustrated in *Figure 4.6*, has been proposed for the chymotrypsin reaction. Histidine-57 acts as a base to activate the serine hydroxyl for nucleophilic attack on the carbonyl carbon of the scissile peptide bond. As the bond is formed between the oxygen of serine and the substrate, the stereochemistry at the carbonyl carbon alters from planar to tetrahedral and charge is dispersed to the carbonyl oxygen (*Figure 4.6b*). The transient charge that accumulates on this oxygen is stabilized by enzymic hydrogen-bonding

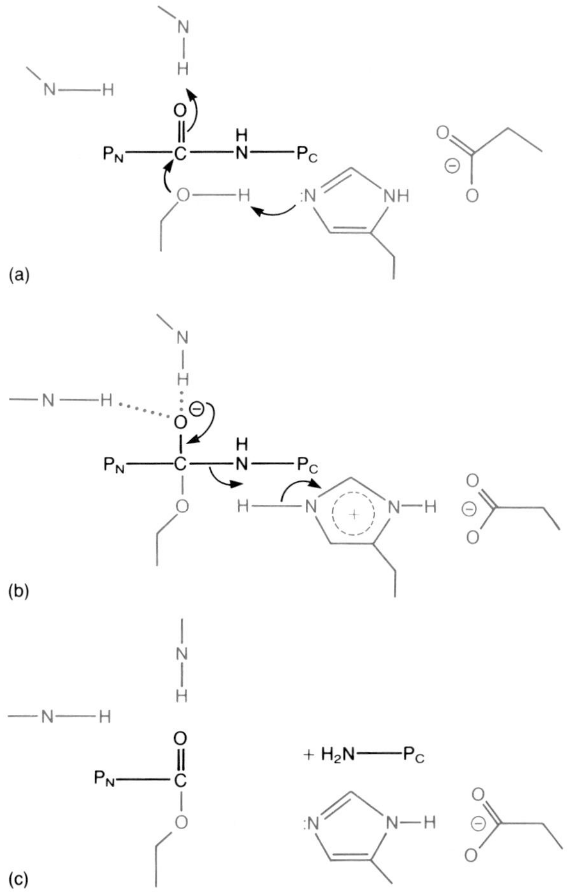

Figure 4.6. The charge–relay mechanism for the formation of the acyl-enzyme intermediate, showing (a) the enzyme–substrate complex, (b) the tetrahedral intermediate, and (c) the acyl-enzyme.

groups. At the same time, the charged histidine-57 interacts with aspartate-102. The transition-state then collapses to the acyl-enzyme (*Figure 4.6c*). Deacylation then takes place as a reverse of these steps, with water replacing serine as the nucleophile and the N-terminal peptide acting as the leaving group.

Despite the wealth of evidence to support the above mechanism, there has been considerable eagerness to test these ideas using site-directed mutagenesis. Developing a system for expression and mutagenesis of serine proteases is challenging because of complexities in their biosynthesis, and the earliest progress was made with the bacterial enzyme, subtilisin.

3.3 Expressing serine protease mutants

To combat the intrinsically destructive activity of the serine proteases, cells synthesize these enzymes as inactive precursors. The precursors are processed to yield the active enzyme only after they have been secreted from the cell or into appropriate vesicles.

Subtilisin is synthesized by *Bacillus* strains of bacteria as the precursor preprosubtilisin. Its activation following secretion involves an autocatalysed proteolysis. There is a problem in expressing mutant subtilisins that have significantly reduced protease activities because these mutants will also have significantly lowered proteolytic activities so that the precursor will accumulate. This problem has been solved in an imaginative way (10). *Bacillus* strains expressing the mutant genes are co-cultured with a small number of cells secreting the active wild-type subtilisin. The active subtilisin will be able to process the mutant precursor. This introduces the complication that it will be necessary to purge the mutant subtilisin preparation of wild-type 'contaminating' activity. Failure to do this will severely compromise any kinetic analysis (Chapter 3). To isolate the mutant subtilisins the system was set up so that each mutant enzyme also has a surface-accessible serine-24 to cysteine-24 mutation. This surface sulphydryl group allows the mutant proteins to bind reversibly, covalently and selectively to thiol Sepharose. Simple column chromatography then allows the purification of the subtilisin mutants.

Expression of recombinant trypsin has also been achieved in a mammalian cell line. The choice of a eukaryotic rather than a bacterial expression system reflects the need to preserve the normal processing, secretion, and disulphide bond-formation factors for this eukaryotic protein.

3.4 The catalytic triad

The involvement of the active site serine-221 and histidine-64 residues in catalysis has been confirmed in subtilisin (SBT) by making alanine replacements (10). The peptidase activities of SBT(Ala 221) and SBT(Ala 64) towards the synthetic substrate, *N*-succinyl-Ala-Ala-Pro-Phe-para-nitroanilide (NsAAPFpNA), were compared with that of the native enzyme. The values of k_{cat}/K_M were reduced by factors of 2×10^6, for both enzymes.

As was hinted earlier, the role of the aspartate-32 residue (asparate-102 in chymotrypsin) is less clear. Although it does not act as a base, its invariance among the serine proteases suggests an important role, possibly in maintaining histidine-64 in the correct orientation and tautomeric form for catalysis. It may also have a role in stabilizing the charge that develops on the histidine in the course of the reaction. Replacement of this aspartate residue with alanine in subtilisin gives an enzyme with a 10^4-fold reduced k_{cat}/K_M, confirming its importance for catalysis. Consistent with a catalytic rather than a substrate binding role, all of this effect is manifested in the lowering of k_{cat}.

The role of the aspartate residue in the catalytic triad of trypsin has also been examined by making an asparagine replacement (11). A similar 10^4-fold decrease in the k_{cat}/K_M is observed. As with the subtilisin mutant, almost all of this reduction is associated with k_{cat}, suggesting strongly that this residue is involved in the catalytic process rather than in substrate binding.

Decreases in activity as large as 10^6-fold force a consideration of the possibility of mistranslation of the mutant gene (12). Errors in protein synthesis are thought to occur with a frequency in the range of 1 in 10^4 to 10^5. Is it possible therefore that the low residual activity of the mutant enzyme preparation is due to a low frequency of misincorporation of the wild-type residue instead of the mutant residue during the translation of the mutant genes? This has been ruled out in experiments that show that the low activity of the mutant enzyme is insensitive to incubation in the presence of phenylmethylsulphonyl fluoride (PMSF). PMSF inhibits the wild-type enzyme stoichiometrically and irreversibly.

Curiously, it was found that the double mutant SBT(Ala 64,Ala 221) is no more impaired in its activity than either the alanine-221 or alanine-64 single mutants, suggesting that removal of either of these residues causes a change in the enzyme mechanism. The basal activity therefore probably reflects catalysis by a mechanism involving direct nucleophilic attack by water or hydroxide on the substrate in a single-step reaction, thereby avoiding the acyl-enzyme intermediate altogether. The activation energy is presumably lowered relative to the uncatalysed reaction by solvation of the oxyanion in the same way as for the wild-type enzyme (see next section).

3.5 Transition-state stabilization

In the proposed mechanism, the carbonyl carbon of the susceptible bond approaches a tetrahedral configuration in the transition state as the bond is formed to the hydroxyl of serine-195. The transition state then collapses to give the tetrahedral intermediate (*Figure 4.6*). A key component of the rate enhancement of serine protease catalysed reactions is believed to be hydrogen-bonding of the developing charge on the oxyanion of the transition state. For the pancreatic serine proteases, the plausible hydrogen-bond donors are the main-chain amide groups of glycine-193 and serine-195 (*Figures 4.5* and *4.6*). The roles of these groups cannot be explored using site-directed mutagenesis. For subtilisin one of the equivalent groups is again provided by the main-chain amide of the catalytic serine-221, whereas the other is provided by the side-chain amide of asparagine-155. The role of this asparagine residue in the oxyanion pocket has been evaluated through the construction of a mutant in which this residue is replaced by a threonine residue. Molecular modelling suggested that in the absence of structural rearrangement, the hydroxyl group of threonine-155 would be 1 Å further away from the oxyanion than the amide nitrogen of asparagine. At 3.5 Å, it would be outside the range needed to form a significant hydrogen bond.

The effect of the mutation is dramatic, with k_{cat}/K_M reduced by a factor of 4000 for SBT(Thr 155), due entirely to a reduction in k_{cat} (13). Once again this suggests that asparagine-155 is not involved in interactions with the substrate in its ground state as K_M is unperturbed by truncation of this residue. Instead, it makes a strong interaction with the substrate in its transition state as the decrease in k_{cat} attests. Replacement of the asparagine by leucine has been carried out by a second group (14) who observed similar but slightly reduced effects (k_{cat} reduced 250-fold). The asparagine–oxyanion interaction appears to stabilize the transition state of the reaction by 3.5–5 kcal/mol, which is close to the energy of interaction of a charge with a dipole measured with tyrosyl-tRNA synthetase. Again, it is seen that an enzyme is realizing favourable interactions selectively in the transition state of the reaction.

Protein engineering studies of serine proteases have provided no surprises; the results serve to buttress existing ideas on their mechanism. Nevertheless, these experiments have been definitive and have allowed the contributions that particular functional groups make to catalysis to be quantitated.

4. Chloramphenicol acetyl transferase

Chloramphenicol acetyl transferase (CAT) is a bacterial enzyme that confers resistance to the antibiotic chloramphenicol (Cm). Chloramphenicol blocks protein biosynthesis by binding to the peptidyl transfer site of the bacterial ribosome. CAT catalyses the transfer of an acetyl group from acetyl–CoA to the C3–O position of Cm (*Figure 4.7*). 3-O–acetyl–Cm is unable to bind to the ribosome.

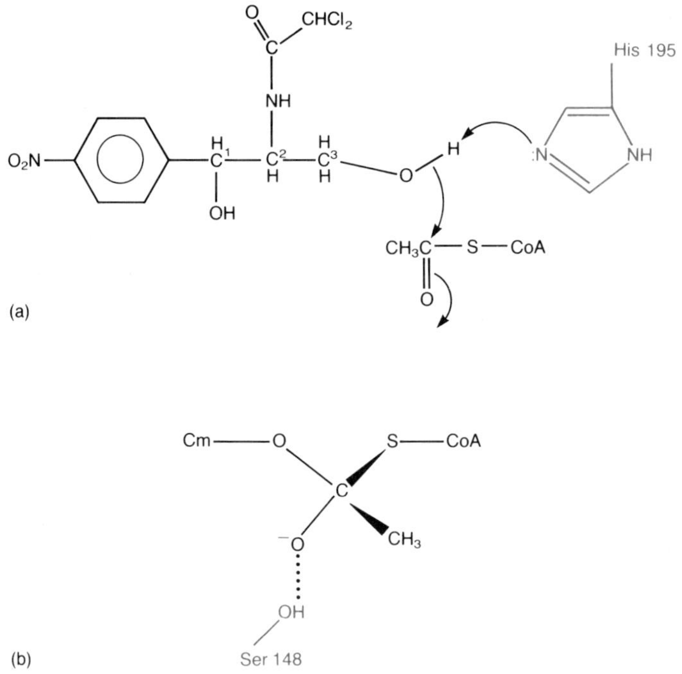

Figure 4.7. Steps in the CAT reaction:
   chloramphenicol + acetyl–CoA ⇌ 3–O–acetylchloramphenicol
(a) General base catalysis by histidine-195 of attack by chloramphenicol C3–O on the thioester of acetyl–CoA.
(b) Interaction of proposed tetrahedral intermediate with serine-148 hydroxyl.

Prior to the elucidation of the structure of CAT (15), the biochemical properties of the enzyme had been studied extensively. Incubation of CAT with the active-site directed reagent, 3-bromoacetyl–Cm, results in alkylation of a conserved histidine-195 at the N_ϵ atom (16). From this and other results, it was suggested that this residue acts as a general base in catalysis. It was further proposed (17) that a nearby acidic residue interacts with the N_δ–H of histidine-195 to assist general base catalysis and stabilize the preferred tautomeric form of the imidazole. This proposal was based on a seductive analogy with the structure and mechanism of serine proteases (Section 3). A prediction of this proposal is that substitution of the acidic side-chain would reduce the basic character of histidine-195, and this would be reflected in a reduction in k_{cat}. Comparison of the amino acid sequences from different sources of the enzyme shows that 27 of the 213 residues are absolutely conserved. Of these conserved residues only two, aspartate-40 and aspartate-199, are acidic. The role of these residues was examined by site-directed mutagenesis to test the hypothesis that an acidic side-chain plays a role in catalysis. Mutation of aspartate-40 to asparagine-40 caused very little change in activity, but CAT(Asn 199) had a considerably reduced k_{cat} (from 599 s^{-1} to 0.28 s^{-1}). The effect on k_{cat} is consistent with a catalytic role for aspartate-199 and appears to support the idea that histidine-195 is activated by this acidic group (18).

When the X-ray structure of the enzyme–chloramphenicol complex was solved, it confirmed that the histidine-195 imidazole makes a hydrogen bond with the C3–O of the Cm (*Figure 4.7a*), consistent with its role as a base in catalysis. However, the orientation of the imidazole ring and its tautomeric stabilization is the result of an interaction with its own carbonyl oxygen (with some steric effects from the nearby tyrosine-25). Aspartate-199 forms a salt-bridge with arginine-18. CAT(Asn 199) crystallized in the same conditions as the wild-type enzyme, but the structure shows significant changes and the histidine ring has moved away from its catalytic position.

The lesson from this is clear, it is possible to attribute effects that are only indirect as being direct when full structural information is unavailable.

A mechanism for the reaction has been proposed (19) whereby the active site histidine-195 acts as a general base, abstracting a proton from the C3 hydroxyl of Cm, thus promoting nucleophilic attack at the carbonyl carbon of the thioester of acetyl–CoA (*Figure 4.7a*). This mechanism predicts the formation of an oxyanion tetrahedral intermediate. The structures of the Cm–CAT and CoA–CAT complexes allowed model building of the proposed tetrahedral intermediate (*Figure 4.7b*). From this model it was proposd that the conserved serine-148 forms a hydrogen-bonding interaction with the oxyanion in the tetrahedral intermediate (19) and is involved in transition state stabilization. A direct prediction of this proposal is that removal of the serine hydroxyl would adversely affect k_{cat}. In CAT(Gly 148), k_{cat} was reduced by only tenfold, but for CAT(Ala 148) k_{cat} was down 58-fold from the wild-type value. The X-ray structure of the CAT(Ala 148) showed that the structural changes were confined to the expected absence of the serine hydroxyl (and some poorly ordered solvent).

The greater activity of the CAT(Gly 148) relative to CAT(Ala 148) can be explained if a water molecule occupies the space created by the removal of the hydroxymethyl group of serine. This water molecule may then partially fulfil the role of the displaced serine hydrogen-bonding group.

The CAT studies underscore the absolute requirement for structural knowledge for the interpretation of enzymological data, and re-emphasize that structure is a fundamental prerequisite for rational protein engineering.

5. Myoglobin: discrimination between O_2 and CO

5.1 Structure and properties

Myoglobin has been the subject of intensive study over the years both as a simple model for protein–ligand interactions and as a baseline for studies of co-operativity in ligand binding, to haemoglobin. It is a muscle protein that binds oxygen reversibly, making it available for aerobic respiration. Myoglobin (Mb) was the first protein crystal structure to be solved, and structures of deoxy as well as oxy (O_2) and carbonmonoxy (CO) forms of the protein have since been solved to high resolution (1.6 Å).

The protein is a monomer that consists of a single 153 amino acid polypeptide chain and a haem cofactor protoporphyrin IX with a co-ordinated iron atom. The three-dimensional structure of myoglobin reveals that the polypeptide chain folds to form eight α-helical segments that are arranged in space to form a hydrophobic pocket in which the haem resides (*Figure 2.1*). In deoxymyoglobin, the iron is co-ordinated to the four pyrrole nitrogens of the porphyrin and to the N_ϵ of the imidazole side-chain of the proximal histidine-93. Oxygen combines reversibly on the opposite, or distal, face of the porphyrin and forms a sixth ligand at the iron (*Figure 4.8*). Throughout, the iron is maintained in the ferrous (Fe^{2+}) state.

Free ferrous haems are also able to bind oxygen; however, they have two properties that render them unsuitable for biological oxygen storage. First, free haem is unable to discriminate against carbon monoxide which it binds some 30 000 times more tightly than oxygen. This is an important consideration because carbon monoxide is produced in the body, paradoxically as a consequence of the breakdown of porphyrins. Second, the haem iron is rapidly oxidized in air to the ferric (Fe^{3+}) state which is unable to combine reversibly with oxygen. The questions of interest are: how does the polypeptide chain modulate the chemistry of the iron–porphyrin group to bind carbon monoxide only 30 times more tightly than oxygen and to slow down the process of haem oxidation from seconds to hours?

5.2 Role of the distal histidine in myoglobin

In the structure of the distal pocket of sperm whale oxymyoglobin, the ligand is in close proximity to the side-chains of three residues, histidine-64, valine-68, and phenylalanine-43 (*Figure 4.8*) (20) which are highly conserved in mammalian myoglobins. A huge amount of research has gone into defining the role of the side-chains in the distal pocket and in particular the role of the distal histidine-64.

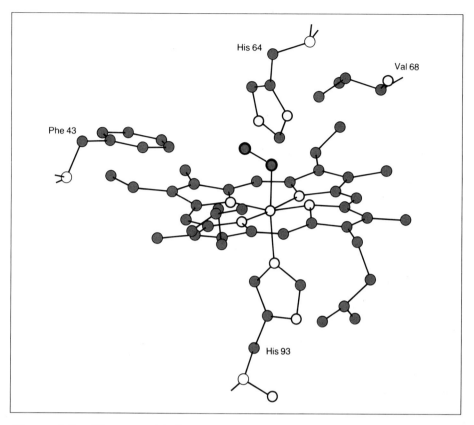

Figure 4.8. The oxygen-binding site in sperm-whale myoglobin showing the residues in the neighbourhood of the bound oxygen.

Neutron diffraction studies of crystals of oxymyoglobin have revealed that there exists a hydrogen bond between the distal histidine N_ϵ and the bound oxygen. The crystal structures of carbonmonoxy complexes of model haem compounds show that the iron–carbon–oxygen bond is linear (180°). In contrast, in the structure of carbonmonoxy–myoglobin, the iron–carbon–oxygen angle is 130°, which is similar to the iron–oxygen–oxygen angle in oxymyoglobin (*Figure 4.8*). It is believed that the distal histidine residue both hinders carbon monoxide from binding with its preferred linear geometry and assists oxygen-binding through the formation of a hydrogen bond with this ligand.

For site-directed mutagenesis studies, a DNA sequence encoding sperm whale myoglobin was chemically synthesized, and spliced downstream of the *lac* promoter for expression in *E. coli*. Mutants at the histidine-64 position were constructed, expressed, purified, and their ligand-binding kinetics were assayed using rapid reaction techniques (21).

The results of the kinetic experiments on mutants at the distal histidine position

Table 4.3. The ligand-binding kinetics for the histidine-64 mutants of myoglobin. The data are taken from ref. 23

Myoglobin	CO			O_2			
	k_1 ($\times 10^{-6}$ s^{-1})	k_{-1} (s^{-1})	K_A ($\times 10^{-6}$ M^{-1})	k_1 ($\times 10^{-6}$ s^{-1})	k_{-1} (s^{-1})	K_A ($\times 10^{-6}$ M^{-1})	M
Mb(His 64)	0.5	0.018	29	16	20	0.8	36
Mb(Gly 64)	5.8	0.038	150	140	1600	0.09	1700
Mb(Val 64)	7.0	0.048	150	250	23 000	0.011	13 600

(*Table 4.3*) support the proposed role of this residue. If the histidine is replaced by a glycine residue in Mb(Gly 64), it is seen that all of the rate constants for the binding and dissociation of oxygen and carbon monoxide are increased. This is expected since the imidazole side-chain presents a physical barrier to ligand entry from the solvent. Its removal is therefore likely to leave a more open distal pocket. For oxygen binding, k_{-1} increases 100-fold whereas k_1 increases only tenfold so that association equilibrium constant, K_A, is decreased by a factor of 10. The decrease in K_A is consistent with the observation that the distal histidine forms a stabilizing hydrogen bond with the bound oxygen. For carbon monoxide, k_1 is increased by a similar factor to that of oxygen; however, the dissociation rate constant is only slightly perturbed, leading to an increase in the association equilibrium constant by a factor of 5. The effect on the discrimination is described by the *M*-value ($M = K_A(CO)/K_A(O_2)$) which is increased 50-fold for Mb(Gly 64).

If the distal histidine is replaced by valine the *M*-value is increased even further, to a value in the range observed for free haems (22, 23). For Mb(Val 64), the steric and hydrogen bonding effects will be similar to those for the glycine mutant. However, whereas the latter may allow water molecules to be accommodated in the distal pocket, the bulkier hydrophobic valine side-chain will cause the environment of the bound ligand to be more hydrophobic. Relative to Mb(Gly 64), Mb(Val 64) has little further effect on CO binding but increases k_{-1} for oxygen by a factor of 15. Unlike bound CO, the bound O_2 ligand is polar and is stabilized by solvation.

Replacing the distal histidine in either Mb(Gly 64) or Mb(Val 64) also causes a 100-fold increase in the rate constant for oxidation of the iron. It is believed that oxidation takes place in the unliganded protein and is promoted by anionic nucleophiles; either hydroxide ions or water molecules. From *Table 4.3*, it can be seen that for the two mutant myoglobins, there will be an increase in the concentration of the unliganded myoglobin species caused by the drop in the oxygen equilibrium association constant. The remaining increase is probably caused by the opening up of the haem pocket by the introduction of the smaller side-chains increasing the accessibility of water and hydroxide ions (23).

6. Repressor proteins

6.1 Introduction

The binding of proteins to specific sequences of DNA is a critical event in the control of gene expression and DNA replication in living systems. These processes in turn, are at the heart of growth and development. An important question is how do these proteins specifically recognize and interact with their target sequences on the DNA in the presence of so many competing non-specific sequences? A significant insight into how this is achieved for one class of DNA-binding proteins has come from studies of the repressor proteins of bacteriophages.

As was mentioned in Chapter 3, the binding of the repressor protein of bacteriophage (phage) λ to its recognition sites on the phage DNA, O_L and O_R, turns off transcription of the phage genes and maintains the lysogenic or quiescent state of the bacteriophage infection. Expression of the repressor protein also serves to confer immunity on the host cell to infection by other phage λ. The binding of repressor to the operator on the incoming phage DNA prevents transcription of the genes whose products are required for the survival and replication of the new phage chromosome. This property of immunity to bacteriophage infection can also be conferred on cells by transforming them with a plasmid carrying a cloned repressor gene.

The determinants of the specificity with which the repressor proteins of two bacteriophages, strains 434 and P22, bind to DNA have been closely examined by Wharton and Ptashne (24). Strain 434 is a λ-like bacteriophage that infects *E. coli*. The 434 repressor recognizes a 14 base-pair DNA sequence of the form

$$5' \text{ ACAAnnnnnnTTGT } 3'$$
$$3' \text{ TGTTnnnnnnAACA } 5'$$

that occurs six times in the 434 phage genome, where n represents a non-consensus base. Phage P22 infects *Salmonella typhimurium*. Its repressor recognizes an 18 base-pair sequence

$$5' \text{ AnTnAAGnnnnCTTnAnT } 3'$$
$$3' \text{ TnAnTTCnnnnGAAnTnA } 3'$$

that occurs six times on the P22 genome.

6.2 Recognition helices

A comparison of the sequences of three DNA-binding proteins whose crystal structures had been determined revealed a set of conserved amino acids (*Figure 4.9*), despite the fact that the three proteins recognize different operator sequences. Significantly, this limited homology was exhibited over a region of the proteins where the backbone structures of the proteins are superimposable. In this region, the structures form a helix-turn-helix motif; that is, an α-helix

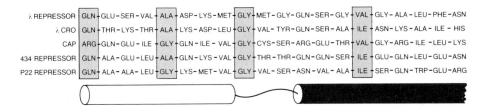

Figure 4.9. Amino acid sequence homologies among five DNA-binding proteins; the shaded residues are conserved. The positions of the amino acids in the helix-turn-helix seen in the crystal structures of λ-repressor, λ-cro, and CAP are indicated beneath the caption. Derived from ref. 24.

followed by a sharp turn followed by another α-helix. Such conservation of structure in proteins with similar DNA binding activities suggests that these motifs have a role in forming the DNA binding domain. In support of this idea model-building indicated that one of the two helices, the recognition helix, could bind to the DNA along its major groove in the manner of *Figure 4.10*.

According to this model the different base sequences in the DNA that are recognized by each of the proteins would be determined by the residues on this α-helix.

6.3 Helix swapping

The conserved sequences described previously are also common to the repressor proteins of phages 434 and P22. It was therefore reasonable to predict that 434 and P22 repressors would also possess the helix-turn-helix motif. If this prediction is correct, then the alignment of the amino acid sequences of the repressor proteins (using the matching residues as a reference) can be used to predict the

Figure 4.10. Schematic diagram showing the interaction of the helix-turn-helix motif of a repressor protein with a DNA duplex. The protein α-helices are represented by cylinders. Derived from ref. 24.

58 Protein Engineering

sequences of the recognition helix. The modelled recognition helices for the two proteins are represented in projection in *Figure 4.11a*. It was presumed that one face of the helix must look inward and interact with the rest of its respective protein while the opposite face is solvent exposed and poised to interact with the specific sequences in the DNA.

To test the 'recognition helix' model, a hybrid protein was constructed by site-directed mutagenesis of the gene encoding 434 repressor. The encoded mutant repressor protein, 434R[αP22] has five substitutions that together constitute a helix-swap. The hybrid repressor possesses a recognition helix that

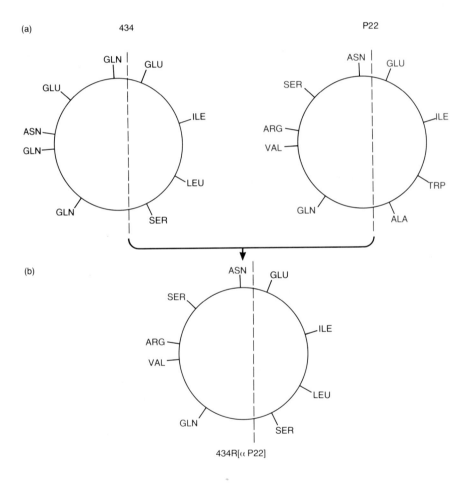

Figure 4.11. (a) Schematic diagram of the recognition helices of phage P22 repressor and of phage 434 repressor. The helices are shown in projection with the side-chain protruding in the directions indicated. (b) Recognition helix in the reshaped protein 434R[αP22].

should, if all the assumptions are valid, outwardly resemble the recognition helix of P22 repressor (*Figure 4.11b*).

The hybrid gene was expressed and the protein assayed for its DNA-binding specificity both *in vivo* and *in vitro*. Cells expressing the mutant repressor 434R[αP22] (i.e. 434 repressor with a reshaped P22 α-helix sequence) from a plasmid were challenged in separate experiments with phage 434 and phage P22. P22 was not able to infect these cells, as would be expected if 434R[αP22] had acquired the ability to bind to P22 operator. In contrast, phage 434 infected and lysed the cells indicating that the hybrid repressor had lost its ability to bind 434 DNA. This interpretation was confirmed by a biochemical analysis of the purified mutant protein. 434R[αP22] binds to purified P22 operator DNA with a very similar affinity to P22 repressor whereas no binding to purified 434 operator sequence was detectable.

These experiments establish that the DNA-binding determinants of the repressors reside in these α-helices, a finding that has now been substantiated by the recent determination of the crystal structure of 434 repressor complexed with a synthetic DNA operator sequence (25).

7. Conclusions

Site-directed mutagenesis has made a profound impact on enzymology. It has allowed penetrating investigation of structure and rigorous testing of hypotheses on mechanism. In many cases, this has simply reinforced existing prejudices; the satisfaction associated with this type of outcome should not be underestimated. In other instances it has forced a serious revision of ideas.

Most importantly, protein engineering has allowed for the first time the measurement of the energetic contributions that the different types of molecular interactions, discussed in Chapter 2, make or can make in aqueous solution. This information will be invaluable in guiding the more ambitious experiments in protein engineering.

8. Further reading

Oxender,D.L. and Fox,C.F. (eds) (1987) *Protein Engineering*. Alan R.Liss, New York.
Fersht,A.R. (1985) *Enzyme Structure and Mechanism*. Freeman, New York.
Leatherbarrow,R.J. and Fersht,A.R. (1986) *Protein Engineering*, **1**, 7.
Knowles,J.R. (1987) *Science*, **236**, 1252.
Shaw,W.V. (1987) *Biochem. J.*, **246**, 1.
Ptashne,M. (1986) *A Genetic Switch: Gene Control in Phage* λ. Blackwell Scientific Publications and Cell Press, Cambridge, MA, USA.

9. References

1. Gardell,S.J., Craik,C.S., Hilvert,D., Urdea,M. and Rutter,W.J. (1985) *Nature*, **317**, 551.
2. Winter,G., Fersht,A.R., Wilkinson,A.J., Zoller,M. and Smith,M. (1982) *Nature*, **299**, 756.

3. Bhat,T.N., Blow,D.M., Brick,P. and Nyborg,J. (1982) *J. Mol. Biol.*, **158**, 699.
4. Fersht,A.R., Shi,J.-P., Knill-Jones,J., Lowe,D.M., Wilkinson,A.J., Blow,D.M., Brick,P., Carter,P., Waye,M.M.Y. and Winter,G. (1985) *Nature*, **314**, 235.
5. Wilkinson,A.J., Fersht,A.R., Blow,D.M. and Winter,G. (1983) *Biochemistry*, **22**, 3581.
6. Leatherbarrow,R.J., Fersht,A.R. and Winter,G. (1985) *Proc. Natl. Acad. Sci. USA*, **82**, 7840.
7. Lowe,G. and Tansley,G. (1984) *Tetrahedron. Lett.*, **40**, 113.
8. Khan,S.A., Kirby,A.J., Wakselman,M., Horning,D.P. and Lawlor,J.M. (1970) *J. Chem. Soc. (B)*, 1182.
9. Blow,D.M., Birktoft,J.J. and Hartley,B.S. (1969) *Nature*, **221**, 337.
10. Carter,P. and Wells,J.A. (1988) *Nature*, **332**, 564.
11. Craik,C.S., Roczniak,S., Largman,C. and Rutter,W.J. (1987) *Science*, **237**, 909.
12. Schimmel,P. (1989) *Acc. Chem. Res.*, **22**, 232.
13. Wells,J.A., Cunningham,B.C., Graycar,T.P. and Estell,D.A. (1986) *Phil. Trans. R. Soc. (Lond.), A*, **317**, 415.
14. Bryan,P., Pantoliano,M.W., Quill,S.G., Hsiao,H.-Y. and Poulos,T. (1986) *Proc. Natl. Acad. Sci. USA*, **83**, 3743.
15. Leslie,A.G.W., Moody,P.C.E. and Shaw,W.V. (1988) *Proc. Natl. Acad. Sci. USA*, **85**, 4133.
16. Kleanthous,C. and Shaw,W.V. (1984) *Biochem. J.*, **150**, 211.
17. Kleanthous,C., Cullis,P.M. and Shaw,W.V. (1985) *Biochemistry*, **24**, 5307.
18. Lewendon,A., Murray,I.A., Kleanthous,C., Cullis,P.M. and Shaw,W.V. (1989) *Biochemistry*, **27**, 7385.
19. Lewendon,A., Murray,I.A., Shaw,W.V., Gibbs,M.R. and Leslie,A.G.W. (1990) *Biochemistry*, **29**, 2075.
20. Phillips,S.E.V. (1980) *J. Mol. Biol.*, **142**, 531.
21. Olson,J.S., Mathews,A.J., Rohlfs,R.J., Springer,B.A., Egeberg,K.D., Sligar,S.G., Tame,J., Renaud,J.-P. and Nagai,K. (1988) *Nature*, **336**, 265.
22. Perutz,M.F. (1989) *Trends Biochem. Sci.*, **14**, 42.
23. Springer,B.A., Egeberg,K.D., Sligar,S.G., Rohlfs,R.J., Mathews,A.J. and Olson,J.S. (1989) *J. Biol. Chem.*, **264**, 3057.
24. Wharton,R.P. and Ptashne,M. (1986) *Trends Biochem. Sci.*, **11**, 71.
25. Aggarwal,A.K., Rodgers,D.W., Drottar,M., Ptashne,M. and Harrison,S.C. (1988) *Science*, **242**, 899.

5

Tailoring protein properties and function

1. Introduction

The previous chapter examined how protein engineering has allowed the dissection of structure–activity relationships in proteins using genetic techniques. This approach has answered long-standing questions of interest to protein chemists and enzymologists and allowed quantitation of some of the energies of interaction of biologically important chemical groups in solution. Undoubtedly, this has reinforced the present understanding as to how proteins work. Here examples of protein engineering are considered; that is, the modification of the structure of a protein to tailor its properties in a predetermined and useful manner. This area is in its infancy at present and progress will be made the faster as a fuller understanding of protein stability and folding emerges.

2. Engineering faster-acting insulins

2.1 Insulin and diabetes

Insulin is a polypeptide hormone whose function is to instruct the tissues to respond to increases in blood glucose concentrations that would, for instance, follow a meal. Insulin exerts its effects through its interaction with insulin receptor molecules which are present on the surface of target tissues. Hormone–receptor binding represents a signal which is transduced across the membrane resulting in changes in cellular metabolism.

The active hormone, the form that interacts with the receptor, is a monomer, which comprises two polypeptide chains A and B of 21 and 30 amino acid residues respectively that are held together by two interchain disulphide bonds.

Insulin is synthesized in the β-cells of the islets of Langerhans in the pancreas as a single polypeptide precursor preproinsulin. The precursor is processed by proteolytic cleavages and stored in the pancreas prior to its secretion into the bloodstream under appropriate physiological circumstances. The biosynthesis and storage of insulin are complex but reasonably well-understood processes that involve oligomerization of the protein.

Diabetes is a common disease often caused by a failure to secrete active insulin. Unless treated, patients become *hyper*glycaemic and would die. It is estimated that there are some 40–80 million diabetics world-wide. Thanks to the pioneering work of Banting and Best in the 1920s diabetes can be controlled by regular injections of exogenous insulin. Pig and beef pancreas have been the major sources of the hormone in the past though now a primary source is human insulin expressed in genetically-engineered *E. coli*.

2.2 Problems of therapeutic insulins

As an effective agent for the treatment of diabetes, insulin is a product of enormous commercial as well as medical importance. A serious deficiency in the preparation and delivery of therapeutic insulins is the time lag between injection of the hormone and its arrival in the bloodstream. This is illustrated in *Figure 5.1*, which compares the profile of the blood insulin concentration following a meal in a non-diabetic individual (A) with that of a diabetic after an insulin injection (B). Insulin cannot be injected directly into the bloodstream for this would have an extreme effect and cause *hypo*glycaemia; rather it is injected into the subcutaneous tissue. The sluggish rate of appearance of insulin in the blood is due to its slow rate of absorption by the tissues. The latter has been attributed to the intrinsic tendency of insulin to form dimers, hexamers, and other aggregates. This self-assembly is an inheritance from the biosynthetic and storage requirements mentioned above. The oligomerization of insulin has been

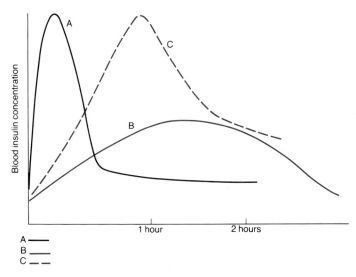

Figure 5.1. Blood insulin profiles: (A) following a meal in a non-diabetic individual, (B) following the injection of purified insulin into the subcutis, and (C) following injection of the engineered aspartate-B28 insulin into the subcutis.

characterized in solution. The molecules dimerize with an association constant (K_A) of 10^6 M^{-1} and the dimers form hexamers with a K_A of 10^9 M^{-2}. Insulin is normally injected at millimolar concentrations so that the monomeric form will represent only a tiny proportion (0.1%) of the total preparation. The oligomerization is an unnecessary obstacle for therapeutic insulin because it slows down release of the hormone and it is not required for its biological effect. Intense efforts have been made to prevent oligomerization by chemically modifying the protein and by varying the composition of the adjuvant with which the hormone is injected. However, the problems remain.

The self-assembly and the activity of insulin are separable, at least notionally. The cloning and expression of the insulin gene in bacteria provides a completely different biosynthetic route to therapeutic insulin; one in which processing, storage, and secretion are not important considerations. For recombinant insulins therefore, oligomerization is redundant, allowing us to explore whether the assembly and activity can be separated physically. Thus, Brange and collaborators (1) set out to engineer active monomeric insulins.

2.3 Protein engineering of monomeric insulins

The strategy adopted for engineering monomeric insulin was to try to disrupt the first stage in the assembly process; the formation of dimers. Essential for this work is the high resolution (1.5 Å) crystal structure of the insulin hexamer (*Figure 5.2a*) (2). The structural model is awe-inspiring in its complexity. Nevertheless, clearly identifiable from the model are the surfaces on the monomers involved in dimer formation. These surfaces are perfectly complementary one with another (*Figure 5.2b*) with the monomer–monomer interaction made up of hydrogen bonds, salt bridges, and van der Waals contacts. The experiments that were devised to upset intermolecular interactions were simplicity itself.

Amino acid substitutions were made in insulin that would either increase steric bulk in a closely packed part of the dimer interface or introduce negatively charged groups into the structure that would oppose existing negatively charged groups in the dimer. Unfortunately, in contrast to the surfaces involved in dimer formation, there is limited information available on the nature and extent of the molecular surface that binds to the insulin receptor. Each mutant insulin therefore had to be assayed not only to determine its aggregation state but also to select those that retained biological activity.

From *Figures 5.3* and *5.2* it is apparent that the valine-B12 residue makes a very snug van der Waals contact with its equivalent in the second molecule of the dimer in a hydrophobic region of structure. Careful analysis of this region of the crystal structure shows that the atoms are closely packed with little or no empty space. To destabilize these interactions, isoleucine was substituted at B12 so that two extra methyl groups would be introduced in the dimer. This conservative substitution was chosen because valine-B12 helps to define the monomer structure and is believed to interact with the receptor; it was hoped that the preservation of the two γ-methyl groups in the side-chain would minimize any losses in the receptor-binding affinity.

64 Protein Engineering

(a)

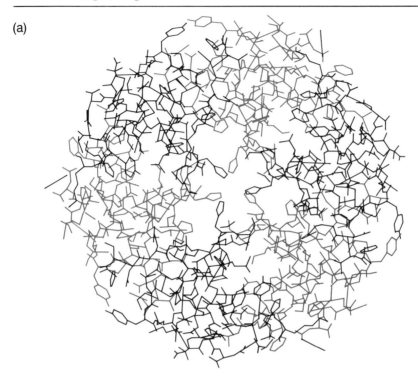

(b)

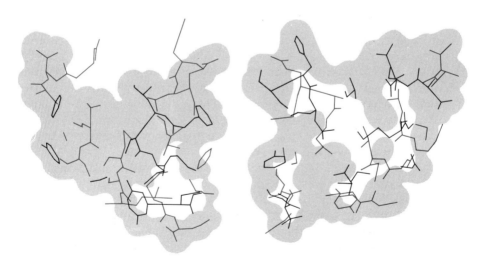

Figure 5.2. (a) Crystal structure of the insulin hexamer. (b) Representation of one of the insulin dimers from the hexamer illustrating the complementary surfaces. In this illustration the monomers have been moved apart by several Ångstroms and a dot surface has been displayed at the van der Waals contact distance from each atom.

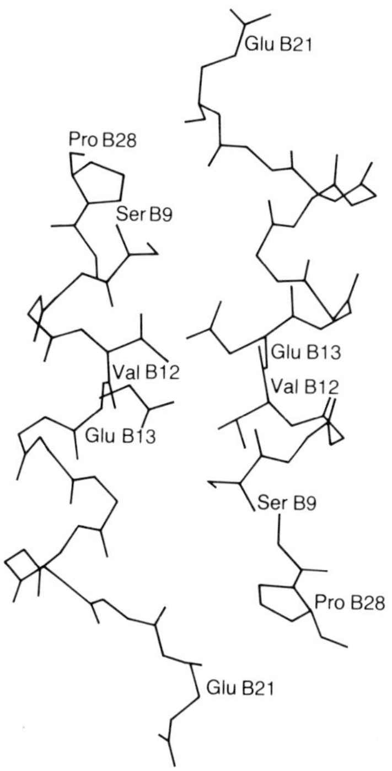

Figure 5.3. Residues at the monomer–monomer interface in the insulin dimer that were targeted in the monomeric insulin engineering experiments.

Surprisingly, isoleucine-B12 was not effective at preventing oligomerization; K_A for dimer formation was decreased by only 30%. Indeed, it was possible to grow crystals of the hexamer, the structure of which reveals that the extra methyl groups are accommodated through a series of structural perturbations of less than 0.5 Å which radiate outwards from the B12 residue in each of the monomers (3). This unexpected result illustrates the plasticity of protein structures and emphasizes how the effects of amino acid substitutions can be propagated beyond the site of their introduction. The biological activity of this mutant insulin was also little affected, showing that any important specific contacts to the receptor made by valine-B12 are preserved with the isoleucine substitution.

In contrast, the introduction of negative charges in the monomer that would oppose existing negative charges in the dimer did prove to be effective at generating monomeric insulin. The reduction in the dimerization constant correlates well with the expected separation of the charges inferred from the crystal structure by modelling the mutations. The valine-B12 to glutamic acid mutation proved particularly effective at producing the monomer because this residue is already in contact with its equivalent in the dimer; as expected, this mutant also exhibited almost no biological activity. Two mutants, proline-B28

(*Figure 5.3*) to aspartate and serine-B9 to aspartate were also monomeric at millimolar concentrations. Aspartate-B28 is expected to be adjacent (4 Å) to glutamate-B21 in the dimer. Likewise, the packing in the dimer would bring aspartate-B9 to within 4 Å of glutamate-B13. Both of these mutants retained activity in standard free fat cell and blood glucose assays, with the aspartate-B28 mutant having the full activity of the wild-type protein.

Most importantly, these two insulins were absorbed twice as rapidly after injection into the subcutaneous tissue in pigs relative to wild-type human insulin preparations. *Figure 5.1* curve C, shows the blood insulin profile following injection of the aspartate-B28 mutant into the subcutis. The curve shows that the mutant insulin appears more rapidly and attains higher concentrations in the blood than the wild-type hormone (1). This augers well both for the treatment of diabetics and for protein engineering.

The production of active monomeric insulin clearly shows that it is possible to use protein crystal structures as a framework for redesigning protein properties, and has demonstrated that there are real prospects for protein engineering. Such studies take protein engineering out of the laboratory and into the everyday world. It is salutary that this simple, even naïve modelling from crystal structures can lead to experiments that may in this case lead to insulin preparations that will improve the therapy for millions of diabetics.

3. Antibodies

3.1 The molecular immune system

The antibodies form a central part of the body's defence against infectious agents and other potentially harmful foreign macromolecules. Their function is to bind to surfaces on the foreign agent (antigen) and trigger its neutralization and elimination, either through stimulating phagocytic cells or through activating the complement-fixing cascade to cause lysis of cells. Their potential for protein engineering arises both from the breadth of structures they can recognize and from the high affinity and specificity they have for their antigens, which is comparable with that of an enzyme for its substrate.

3.2 Immunoglobulin structure

The immunoglobulin molecule is made up of four polypeptide chains consisting of two identical light chains (of about 220 amino acids) and two identical heavy chains (of about 440 amino acids). Immunoglobulin molecules bind specifically to their antigen and this binding serves to trigger effector functions. There are distinctive classes of heavy chain which are associated with a distinctive set of effector functions. The basic structure of the antibody is a Y-shaped molecule with two antigen combining sites as indicated in *Figure 5.4a*.

The antigen binding surface of the antibody is formed by sequences at the N-termini of both the heavy and light chains. It is variability among these sequences that endows antibodies with their different antigen recognition properties. Closer analysis of these variable regions shows that they comprise conserved sequences that are interspaced with hypervariable regions (*Figure 5.4a*). Crystal structures of antibody–antigen complexes reveal clearly that the six hypervariable regions cluster to form the antigen combining surface. The conserved sequences within the variable region provide a rigid framework for the hypervariable loops that generate antibody diversity. The F_c region of the antibody (*Figure 5.4a*) is responsible for triggering the necessary effector functions that follow the combination of the antibody with the antigen.

3.3 Antibody therapy

Antibodies have enormous potential as therapeutic agents. They can be administered to patients to treat diseases that could prove lethal before an effective immune response can be mounted. They would also be particularly suitable for treating patients with immune deficiencies. Antibodies against tumour cells may also prove a route to cancer therapy either through their natural physiological effector activating mechanisms, or alternatively the antibody might be conjugated to toxins or radionuclides and used to deliver a 'magic bullet'.

Antibody secreting cells proliferate after mammals have been challenged with antigen. Mice and rats have provided a convenient experimental system for raising antibodies against potentially toxic or pathogenic antigens. Moreover, it is a routine procedure to isolate and immortalize individual antibody secreting

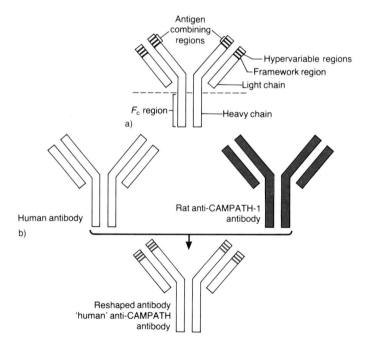

Figure 5.4. (a) Schematic structure of an antibody. (b) Reshaping a human antibody with hypervariable loops from a rat antibody.

cells (4) from rats or mice, and hence grow monoclonal cell cultures that secrete a single distinct antibody. Practical as well as ethical considerations mean that most antibodies are raised in rats or mice.

For treating human patients, however, it is highly desirable to use human antibodies. This is because non-human antibodies will be recognized as foreign and the human immune system will most likely mount a response against them. Furthermore, it is expected that human antibodies will be better able to activate human effector responses.

3.4 Reshaping human antibodies

There has been considerable interest in 'humanizing' rat or mouse antibodies. This has been done by splicing together the variable regions of the heavy- and light-chain rat genes with the corresponding constant regions of the human genes (5). Following insertion of these genes into expressing cell lines, chimaeric antibodies can be isolated. These antibodies have the antigen combining surfaces of the rat antibody but are otherwise human. Because they contain far less foreign sequence than the rat antibody they should be less immunogenic in humans.

This philosophy has been extended to the 'grafting' of the hypervariable loops alone, from the mouse or rat antibodies on to the framework and constant regions of the human antibody (6, 7). Six sequences in the DNA, three each in the heavy- and light-chain genes, must be modified. Studies by Winter and colleagues have focused on a promising rat antibody that is directed against an antigen known as CAMPATH-1. The occurrence of this antigen is restricted to the surface of lymphoid cells and monocytes where it is abundantly expressed. CAMPATH-1 has been suggested as a potential target for antibodies to combat a cancer, caused by the uncontrolled proliferation of these cells, called non-Hodgkin lymphoma.

The hypervariable sequences of the anti-CAMPATH-1 antibody were determined by DNA sequencing and long oligonucleotides were prepared that would direct the incorporation of these rat hypervariable region coding sequences into the heavy- and light-chain genes of a human antibody (*Figure 5.4b*). These genes were then introduced into a cell-line and the reshaped antibodies were expressed. Preliminary assays showed that the humanized antibody bound effectively to the CAMPATH antigen and that this binding activated complement and cell-mediated lysis. Clinical trials with the genetically engineered antibody have so far proved very promising (8). In patients treated with daily doses of the antibody over a period of several weeks, lymphoma cells were efficiently cleared from the blood and bone-marrow without any anti-immunoglobulin response being mounted.

3.5 Catalytic antibodies

Antibodies also provide an interesting alternative route for generating novel protein catalysts. The strategy is based on the concept that enzymes have evolved to be complementary to the transition states of their substrates. It is reasoned therefore that if an antibody can be selected that binds tightly a transition state analogue for a particular reaction, it may possess catalytic properties (9). Such experiments have met with some, albeit modest, success and work is continuing. One particular attraction of this approach is that it does not require a detailed consideration of folding. A second is that it provides an approach to obtaining catalysts for reactions that need have no relation to existing biochemical processes.

A slightly different and more ambitious approach is to isolate an antibody against the substrate for a particular reaction and then to use site-directed mutagenesis to introduce the catalytic groups.

4. Enhancing activity in tyrosyl-tRNA synthetase

The conclusions from the studies of substituting hydrogen-bonding residues in tyrosyl-tRNA synthetase of *B. stearothermophilus* (Section 2 of Chapter 4) suggest that long hydrogen bonds are weak and contribute little to catalysis. Deletion of such hydrogen-bonding groups (e.g. serine-35 to glycine, *Table 4.1*) has little or no deleterious consequences for catalysis and may even result in small increases in activity. An inspection of the crystal structure of the enzyme – intermediate complex (*Figure 4.1*) reveals one possible hydrogen bond (between threonine-51 and the ring oxygen of the ribose) that is longer than is optimal (3.5 Å instead of 2.8 Å). When threonine-51 is substituted by alanine-51 the data in *Table 5.1* show that the enzyme activity in the activation reaction is enhanced for all concentrations of the substrate, ATP (10). The enzyme substrate binding energy in the transition state is increased by 0.4 kcal/mol.

When this residue is changed to a proline, which is found at the equivalent position in the homologous *E. coli* enzyme, there is a dramatic increase in enzyme activity; the transition states for both the activation and aminoacylation reactions are stabilized by 1.9 and 2.3 kcal/mol respectively relative to the wild-type enzyme (*Table 5.1*). TyrRS(Pro 51) is a significantly more efficient catalyst of the activation reaction at all substrate concentrations. Examination of the

Table 5.1. Activation and aminoacylation data for the position-51 mutants of TyrRS at 298K (10).

Enzyme	k_{cat} (s^{-1})	K_M (ATP) (mM)	k_{cat}/K_M $(s^{-1} M^{-1})$	$\Delta\Delta G_T^{\ddagger}$ (kcal/mol)
Activation				
TyrRS(Thr 51)	7.6	0.9	8400	–
TyrRS(Ala 51)	8.6	0.54	15 900	0.38
TyrRS(Pro 51)	12.0	0.058	208 000	1.9
Aminoacylation				
TyrRS(Thr 51)	4.7	2.5	1860	–
TyrRS(Ala 51)	4.0	1.2	3200	0.32
TyrRS(Pro 51)	1.8	0.019	95 800	2.33

aminoacylation data, however, shows that although k_{cat}/K_M is increased 50-fold, this comes about through a 130-fold reduction in K_M and a small two- to three-fold reduction in k_{cat}. For this reason the proline-51 enzyme is a more efficient catalyst of aminoacylation only at very low substrate concentrations. At physiolgical ATP concentrations, in the millimolar range, TyrRS(Pro51) is saturated with ATP so that the rate is controlled by k_{cat} rather than k_{cat}/K_M. This confers a large disadvantage on TyrRS(Pro 51).

Another way to evaluate this is to re-examine Equation 3.9, which shows that although rate increases as k_{cat}/K_M increases it also decreases as $[E_f]$ decreases. As has been shown in Equation 3.8, K_M is an apparent dissociation constant for all the enzyme-bound species; $K_M = [E_f][S]/\Sigma[ES]$. Decreasing K_M below the substrate concentration encountered *in vivo* therefore militates against catalysis because the enzyme becomes sequestered in the enzyme–substrate complex and the concentration of free enzyme, $[E_f]$, is reduced.

This example illustrates the constraints that are imposed on an enzyme in the course of evolution. The enzyme must bind the transition state as tightly as possible to bring down the highest energy level on the reaction co-ordinate. However, the tight binding of the transition state often necessarily involves tighter binding of the ground state. This becomes undesirable when the substrate is bound so tightly that the enzyme becomes sequestered in the enzyme–substrate complex (*Figure 5.5*). This is what happens in the case of TyrRS(Pro 51), where the enzyme and substrate fall into a thermodynamic 'pit' on forming the enzyme-substrate complex. From this 'pit' there is a large energy barrier to attaining the transition state ES$^\neq$ (*Figure 5.5*). Therefore, the wild type enzyme

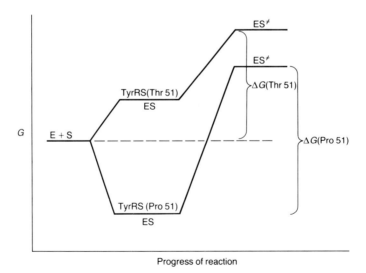

Figure 5.5. Energy profiles for tyrosyl-adenylate formation catalysed by TyrRS(Pro 51) and TyrRS(Thr 51) at millimolar concentrations of ATP. The largest barrier (peak to trough) is greater for the mutant ΔG(Pro 51) than it is for the wild-type enzyme ΔG(Thr 51).

has evolved to bind most tightly those structures that are unique to the transition state (11, 12). This is achieved by maximizing k_{cat}/K_M without reducing K_M below the substrate concentration found *in vivo*. In so doing, the enzyme minimizes the energy difference between the lowest trough and the highest crest on the reaction co-ordinate which facilitates catalysis.

5. Subtilisin: practical applications

5.1 Overview

The properties demanded of an enzyme to be used in an industrial process or in a commercial product are usually very different from those imposed on the enzyme through evolution. The most obvious difference between these needs is that demanded because the enzymes are working in different environmental conditions. These may include the temperature, pH, ionic strength, dielectric constant, and redox potential, as well as the presence or absence of inhibitors, detergents, and denaturants. Considerable effort is currently being directed to honing the physical properties of enzymes such as amylases, lipases, and proteases to improve and expand their industrial usefulness.

Subtilisin is an enzyme of considerable commercial importance; it is an important component of washing powders. Subtilisin catalysis has been discussed previously (Section 3 of Chapter 4); here, experiments in which the stability, pH profile, and specificity of bacterial serine proteases have been modifed are briefly discussed.

5.2 Engineering resistance to chemical oxidation

One of the mechanisms by which enzymes become inactivated is through chemical oxidation. Particularly susceptible sites are at the sulphur-containing residues cysteine and methionine. Subtilisin is inactivated in the presence of hydrogen peroxide. The mechanism of inactivation is through the formation of a sulphoxide at the invariant methionine-222 residue.

A very obvious route to engineering oxidative stability is simply to replace this residue in a site-directed mutagenesis experiment. Because there were uncertainties *a priori* as to which substitution would be the most suitable from the point of view of preserving stability and activity, mutagenesis was used systematically to introduce all 19 amino acid replacements (13).

The analysis of the mutant proteins showed that with the exception of the cysteine-222 enzyme all of the mutants had decreased specific activities in the hydrolysis of peptide substrates, though for a number of these including the serine-222 and alanine-222 enzymes, the decrease was only by a factor of two to three. Assays performed after incubation of SBT(Ser 222) and SBT(Ala 222) in the presence of 1 M hydrogen peroxide for 15 minutes revealed no loss of activity. In similar assays of wild-type enzyme the activity dropped to 10% of its initial value after 2 minutes. Clearly, the introduction of increased oxidative stability has been achieved.

5.3 Modification of the pH/activity profile

The pH-dependence of the reaction catalysed by subtilisin has been altered by changing the surface charge on the enzyme. Subtilisin activity is lost upon protonation of the active site histidine-64 which titrates with a pK_a of 7.2. The pK_as of ionizing groups in proteins are influenced by their immediate environment. Indeed, the pK_a of histidine-64 is higher than that of a free imidazole in solution because of the proximity of the charged side-chain of aspartate-32.

Thomas et al. (14) have replaced an aspartate residue (aspartate-99) which is on the surface of the protein some 14 Å from the active site with a serine. The pH-dependence of k_{cat}/K_M for the hydrolysis of sAAPFpNA by the aspartate-99 and serine-99 proteins in a buffer with an ionic strength of 0.1 is illustrated in Figure 5.6. The figure shows that the pH/activity profile for the protease has been shifted for SBT(Ser 99), so that the pK_a of the histidine-64 is 6.9. That the perturbation of the pK_a is caused by an electrostatic effect was confirmed by repeating these experiments in a buffer with an ionic strength of 1.0. At high ionic strength electrostatic effects are masked, and the serine-99 mutation has essentially no effect.

Subsequent studies on subtilisin indicate that the introduction of further mutations, that remove negatively charged groups or introduce positively charged groups elsewhere on the protein surface, produce enzymes with greater decreases, of up to 1.0 pH unit, in the pK_a of the histidine (15). This approach would appear to have general applicability for tailoring the pH-dependence of enzyme catalysis. Furthermore, as these mutations involve surface residues that can be up to 15 Å from the active site, it is expected that protein folding and stability will be minimally affected.

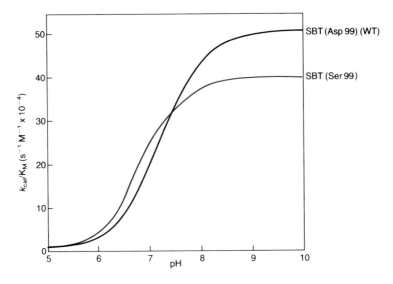

Figure 5.6. Altering the pH/activity profile in subtilisin.

5.4 Altering substrate specificity

The configuration of the catalytic groups is identical in the crystal structures of a number of serine proteases. However, this conservation of structure does not extend to the P_1 site where specificity is conferred by means of alterations in the size, polarity and charge contained within this pocket. The P_1 site on the enzyme is where the group R in Equation 4.9 is bound. Thus, for chymotrypsin a large hydrophobic pocket accommodates the aromatic side-chains while for trypsin there is an aspartate side-chain that can form ionic interactions with the large, positively charged side-chains of lysine and arginine.

The specificity of bacterial serine proteases has been altered through substituting residues in the P_1 pocket. α-Lytic protease (αLP), like subtilisin, is an extracellular protease. Whereas subtilisin prefers large hydrophobic substrates, αLP has a preference for substrates that allow small hydrophobic side-chains to bind at the P_1 position. This can be seen from the activity of the enzyme towards different substrates of the form N-succinyl-Ala-Ala-Pro-X-para-nitroanilide (NsAAPXpNA), where X represents the amino acids alanine, methionine, or phenylalanine (*Figure 5.7*). The X-ray structure shows that one of the residues lining the P_1 pocket is methionine-192, which occludes what would otherwise be a large hydrophobic pocket. Methionine-192 has been replaced by alanine in αLP(Ala 192) (16). In the absence of compensating movements of neighbouring residues, such a mutation would be expected to expand the hydrophobic pocket, possibly accommodating larger side-chains. The data in *Figure 5.7* show that the preference of α-lytic protease for the alanine substrate over the methionine substrate is reversed through the methionine to alanine substitution. Quite astonishingly, there is an enormous 10^5- to 10^6-fold

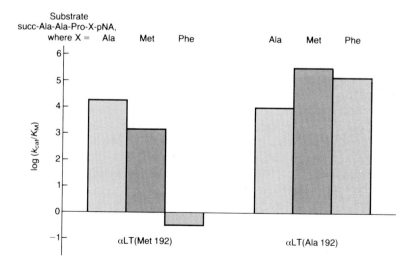

Figure 5.7. The specificities of wild-type α-lytic protease, αLP(Met 192), and αLP(Ala 192) towards NsAAPXpNA substrates.

increase in the activity of the αLP(Ala 192) towards the phenylalanine substrate relative to the wild-type enzyme.

In contrast, the specificity of subtilisin has been modified in the direction of smaller amino acids by the introduction of larger hydrophobic side chains at glycine-166 (17) which reduce the volume of the P_1 pocket.

These experiments illustrate that it is possible to think of the enzyme-substrate binding surfaces as being complimentary. Steric bulk has been shuffled from the enzyme to the substrate to influence the specificity in a predictable manner. Impressive alterations in the substrate specificity of lactate dehydrogenase, lactate to malate (18), and glutathione reductase, $NADP^+$ to NAD^+ (19), have been achieved by using similar simple mutagenesis strategies.

6. Engineering thermostability in lysozyme

6.1 Protein stability

There is much scope in protein engineering for the design and preparation of proteins with increased stability, and this is one of the major goals of the field. The problem in understanding stability arises from the large number and types of non-covalent interactions that are made in both the folded and unfolded structures of a protein. Together these individual terms add up to thousands of kilocalories. Stability results from the difference in the free energies of the folded and unfolded states of the protein. The observation from many proteins is that the folded state of the protein is stabilized by as little as 5–20 kcal/mol relative to the unfolded state. This accounts for the characteristic that above physiological temperatures many proteins are unstable with respect to unfolding.

Intuitively, the most important forces stabilizing the folded protein are likely to be hydrophobic interactions. This is because polar interactions, either hydrogen-bonding or electrostatic, are also satisfied in the unfolded and presumably solvent-exposed state. Proteins fold so as to preserve as many of these polar interactions as possible either by keeping these groups exposed to solvent on the outside of the protein or by extensive intramolecular hydrogen-bonding. In contrast, the burying of hydrophobic groups in the apolar interior of a protein is entropically favoured, as this process restores the more disordered hydrogen-bonding arrangement of the solvent which was hitherto interrupted by the apolar hydrophobic side-chains. Opposing this are entropy losses that arise as the many protein conformations in the unfolded state become restricted in the folded molecule.

Bacteriophage T4 lysozyme has proved a fertile ground for exploring and engineering thermostability. The enzyme catalyses the hydrolysis of bacterial cell wall polysaccharides leading to lysis of the cells. The enzyme is a monomer of 164 amino acids that requires no cofactor and possesses no disulphide bridges. The high resolution crystal structure of the enzyme reveals eleven α-helical segments that fold into two domains (*Figure 5.8*) between which there exists a cleft in which the substrate binds. The enzyme unfolds reversibly, and this process

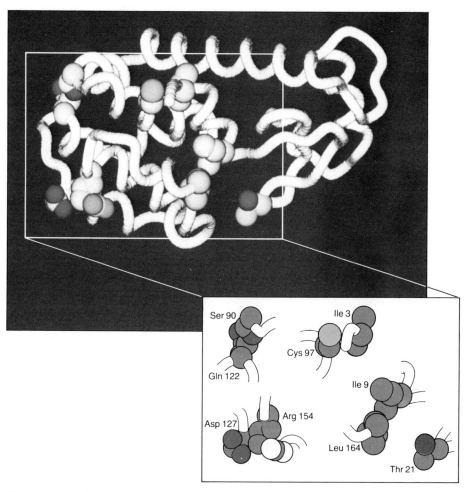

Figure 5.8. The backbone of T4 lysozyme illustrating the positions of the existing cysteine residues and the residues which were mutated to cysteines to introduce disulphide bridges.

can be followed spectroscopically. The unfolding process can be described by the melting temperature T_m, which is the temperature at which the free energy of unfolding is zero. Under these conditions, the equilibrium constant for the interconversion of the folded (F) and unfolded (U) proteins K_U in Equation 5.2 is unity and F and U are present at equal concentrations

$$F \overset{K_U}{\rightleftharpoons} U \qquad (5.2)$$

The stability of a protein results from the difference between the free energy of the folded state and the free energy of the unfolded state. These free energy differences (ΔG) themselves are the sum of enthalpic (ΔH) and entropic ($-T\Delta S$) terms. Ideally, it would be possible to calculate the enthalpic and entropic contributions of individual residues in the protein to the folded and unfolded states and assess the effects of point mutations on stability theoretically. Stumbling blocks to such a calculation include the absence of accurate information about the precise structure(s) of the unfolded protein and a poor understanding of the entropy terms involved. Furthermore, because there are such a large number of balanced terms in the calculation, it will be impossible to obtain the accuracy that is needed (+/− 5 kcal/mol).

6.2 Conformational entropy

Despite the complexities it is nevertheless possible to devise rational approaches to increasing stability. The permissible conformations of the main-chain atoms of the amino acids are described by the Ramachandran plot (*Figure 2.4*). Two amino acids, glycine and proline, have anomalous plots. Glycine has an unusually broad range of allowed values for ϕ and ψ because glycine has a simple hydrogen atom side-chain, whereas the rest of the amino acids have side-chains that possess a bulkier β-carbon atom. In contrast proline has an unusually restricted range of ϕ and ψ values because the α-imino group is locked in the pyrrolidine ring. If the unfolded protein is considered as an open chain in which each residue samples its allowed conformational states and the folded protein to be a more rigid structure in which only a few of these states are sampled, then each residue loses entropy on folding. Residues such as glycine that have a lot of conformational entropy in the unfolded state have more to lose upon folding. Similarly a residue such as proline that is already constrained has less to lose.

Thus a route to increasing protein stability can be reasoned that involves destabilizing the unfolded state of the protein by either increasing the proline content, or decreasing the glycine content of the protein. This approach has been applied to T4 lysozyme (20). The crystal structure was inspected for the existence of residues that matched certain criteria. Glycine residues were identified that have ϕ and ψ values in the range allowable for an alanine residue; of these residues glycine-77 was selected because modelling suggested that an alanine replacement can be made at this position without the β-methyl group interfering sterically with neighbouring residues. For a second experiment, the whole structure was scrutinized to pick out those residues, other than proline itself, with ϕ and ψ values in the range allowable for proline. Alanine-82 was chosen as a residue that would tolerate substitution by proline, without serious structural perturbation.

Thermodynamic studies of the alanine-77 and proline-82 proteins revealed that T_m for both proteins was increased at pH 6.5 by 1° and 2°C respectively. In each case the expected changes in the entropy contribution were observed. Although the changes observed are small, significantly they are in the direction of increased thermostability. Structural studies of these two mutants show that the backbone structures of the proteins are not perturbed by these mutations, supporting the interpretation of the results.

6.3 α-Helix dipole interactions

α-Helices are common elements of secondary structure in proteins. The parallel alignment of the dipoles of the peptide bonds along the direction of the helix axis gives the helix a significant overall dipole moment. The N-terminus of the helix is believed to possess a net positive charge as high as 0.5 electrons, while the C-terminus has a corresponding net negative charge. Consistent with this notion is the asymmetric distribution of charged residues in α-helices. Aspartate and glutamate residues are preferred as N-terminal residues, while lysine, arginine, and histidine tend to be located at the C-termini of helices (21). In each case, the charged side-chain can make favourable electrostatic interactions with the helix dipole.

Inspection of the eleven α-helices in bacteriophage T4 lysozyme led to the identification of two helices as candidates that could accommodate aspartic acid replacements at their N-termini (22). Three mutant proteins were constructed: serine-38 to aspartate, asparagine-144 to aspartate, and a double mutant comprising both of these changes. The melting temperatures of these mutant proteins were measured in the range of pH 2–7 and compared to those of the wild-type enzyme. At pH 2 all three mutant enzymes showed no differences from the wild-type enzyme. In the range of pH 3–7 where the aspartates ionize, the single mutants showed increases in T_m as high as 2°C. This indicates that the charged forms of the aspartates are the important species. As might be expected, the effects of the two mutations were additive, indicating that it may be possible to design a very much more stable enzyme through the combination of a series of mutations that individually give only small increases in stability. The double mutant's T_m was increased by up to 4°C corresponding to a free energy of stabilization relative to the wild-type enzyme of 1.6 kcal/mol (22). Although the magnitudes of the changes in T_m are small, it is again noteworthy that such an apparently naïve approach is effective at all.

6.4 Introducing covalent crosslinks
A third strategy for stabilizing proteins is to introduce covalent crosslinks between residues which are neighbours in the folded structure of the protein. The effect of such a modification will be to tether together these residues in the unfolded state of the protein too, thereby decreasing the entropy of the unfolded state. This approach has been used successfully to stabilize hen-eggwhite lysozyme through a covalent linkage of glycine-35 and tryptophan-108 by chemical means (23). The melting temperature of the modified enzyme at pH 2.0 was increased by 29°C.

A more common form of crosslinking residues in proteins is through disulphide bonds between cysteine residues. These may be engineered into proteins by introducing pairs of cysteine residues at appropriate positions. Five different disulphide bonds have been successfully engineered into T4 lysozyme by introducing cysteines at positions 3 and 97, 9 and 164, 21 and 164, 127 and 154, and 90 and 122 (*Figure 5.8*) (24). Thermodynamic analyses of the oxidized forms of these proteins, in which the —S—S— bridges are formed, showed that the melting temperatures at pH 2.0, for the three mutants' linking positions 3–97, 9–164 and 21–164 were increased by between 5° and 11°C. For the other two proteins, with links at 127–154 and 90–122, the values of T_m were down by 0.5–2°C. Further analysis confirmed that the increased stability arises from a destabilizing entropic effect on the unfolded protein which is partly offset by a destabilizing strain energy on the folded form.

7. Conclusions

It is clear that significant progress has been made in protein engineering in a short period of time. The topics discussed in this chapter show that useful advances can be made in tailoring protein properties and function from a combination of clear scientific reasoning and accurate structural information.

The general process of engineering proteins remains, however, a process of informed trial and error. The intense research activity currently taking place in this area is contributing to an increased pool of data and experience. It can therefore confidently be forecast that protein engineering will continue to yield further insights into protein form and function.

8. Further reading

Brange,J. (1987) *Galenics of Insulin.* Springer-Verlag, Berlin and Heidelberg.
Leslie,R.D.G. (ed.) (1989) *Diabetes, British Medical Bulletin 45.* Churchill Livingstone, Edinburgh.
Roitt,I. (1988) *Essential Immunology.* Blackwell Scientific Publications, Oxford.
Alberts,B., Bray,D., Lewis,J., Raff,M., Roberts,K. and Watson,J.D. (1983) *The Molecular Biology of the Cell.* Garland Publishing, New York and London.
Alber,T. (1989) *Annu. Rev. Biochem.,* **58**, 765.
Matsumura,M., Becktel,W.J. and Matthews,B.W. (1988) *Nature,* **334**, 406.
Serrano,L. and Fersht,A.R. (1989) *Nature,* **342**, 296.

9. References

1. Brange,J., Ribel,U., Hansen,J.F., Dodson,G., Hansen,M.T., Havelund,S., Melburg,S.G., Norris,F., Norris,K., Snel,L., Sorenson,A.R. and Voigt,H.O. (1988) *Nature,* **333**, 679.
2. Baker,E.N., Blundell,T.L., Cutfield,J.F., Cutfield,S.M., Dodson,E.J., Dodson,G.G., Crowfoot Hodgkin,D.M., Hubbard,R.E., Isaacs,N.W., Reynolds,C.D., Sakabe,K., Sakabe,N. and Vijayan,N.M. (1988) *Phil. Trans. R. Soc. (Lond.) B,* **319**, 369.
3. Derewenda,U., Derewenda,Z., Dodson,G. and Brange,J. (1987) *Protein Engineering,* **1**, 238.
4. Köhler,G. and Milstein,C. (1975) *Nature,* **256**, 495.
5. Morrison,S.L., Johnson,M.J., Herzenberg,S.A. and Oi,V.T. (1984) *Proc. Natl. Acad. Sci. USA,* **81**, 6581.
6. Riechmann,L., Clark,M., Waldmann,H. and Winter,G. (1988) *Nature,* **332**, 323.
7. Verhoeyen,M., Milstein,C. and Winter,G. (1988) *Science,* **239**, 1534.
8. Hale,G., Clark,M.R., Marcus,R., Winter,G., Dyer,M.J.S., Phillips,J.M., Riechmann, L. and Waldmann,H. (1988) *Lancet,* **ii**, 1394.

9. Powell,M.J. and Hansen,D.E. (1989) *Protein Engineering*, **3**, 69.
10. Wilkinson,A.J., Fersht,A.R., Blow,D.M., Carter,P. and Winter,G. (1984) *Nature*, **307**, 187.
11. Fersht,A.R. (1985) *Enzyme Structure and Mechanism*. Freeman, New York.
12. Albery,W.J. and Knowles,J.R. (1977) *Angewandte Chemie*, **16**, 285.
13. Estell,D.A., Graycar,T.P. and Wells,J.A. (1985) *J. Biol. Chem.*, **260**, 6518.
14. Thomas,P.G., Russell,A.J. and Fersht,A.R. (1985) *Nature*, **318**, 375.
15. Russell,A.J. and Fersht,A.R. (1987) *Nature*, **328**, 496.
16. Bone,R., Silen,J.L. and Agard,D.A. (1989) *Nature*, **339**, 191.
17. Estell,D.A., Graycar,T.P., Miller,J.V., Powers,D.B., Burnier,J.P., Ng,P.G. and Wells,J.A. (1986) *Science*, **233**, 659.
18. Wilks,H.M., Hart,K.W., Feeney,R., Dunn,C.R., Muirhead,H., Chia,W.N., Barstow,D.A., Atkinson,T., Clarke,A.R. and Holbrook,J.J. (1988) *Science*, **242**, 1541.
19. Scrutton,N.S., Berry,A. and Perham,R.N. (1990) *Nature*, **343**, 38.
20. Matthews,B.W., Nicholson,H. and Becktel,W.J. (1987) *Proc. Natl. Acad. Sci. USA*, **84**, 6663.
21. Richardson,J.S. and Richardson,D.C. (1988) *Science*, **240**, 1648.
22. Nicholson,H., Becktel,W.J. and Matthews,B.W. (1988) *Nature*, **336**, 651.
23. Johnson,R.E., Adams,P. and Rupley,J.A. (1978) *Biochemistry*, **17**, 1479.
24. Matsumura,M., Becktel,W.J., Levitt,M. and Matthews,B.W. (1989) *Proc. Natl. Acad. Sci. USA*, **86**, 6562.

Glossary

Dipole: The electrostatic moment that results from the separation of partial charges within a molecule. These dipoles can interact with other dipoles or charges electrostatically.

Electrophile: Electron seeker. An atom that is deficient in electrons or has a positive or partial positive charge. Its characteristic is its reactivity towards electron-rich (nucleophilic) regions.

Entropy: Disorder. A state with less order is preferred over that with increased order. Entropy (S) is measured in units of kcal mol^{-1} K^{-1}.

Enthalpy: The energy given out or taken up in a chemical process (assuming there is no change in the volume). Enthalpy (H) is measured in kcal mol^{-1}.

Eukaryote: An organism whose cells have discrete nuclei. All higher organisms are in this class.

Hydrogen bond: Local electrostatic interaction that allows a hydrogen atom with a partial positive charge to be shared between two electronegative atoms (e.g. O, N, S). This results in a closer approach than would be allowed by van der Waals radii. Hydrogen bonds between nitrogens and oxygens are typically 2.7–3.3 Å long.

Hydrophilic: A chemical species that has an energetic preference for a polar environment. Polar groups are in this class.

Hydrophobic: Water hating, applied to chemical species that have an energetic preference for non-aqueous environments. Non-polar groups are in this class.

Ion pair: Local electrostatic interaction between charged species of opposite sign.

Nucleophile: The complement of an electrophile. An atom or region of a molecule that is electron rich, and therefore reactive toward electron deficient species.

Prokaryote: An organism that does not have a separate subcellular compartment (nucleus) for its hereditary material. All bacteria are in this class.

Scissile bond: The bond that is susceptible to cleavage during the course of a reaction.

Tautomer: One of the states a molecule can have by the rearrangement of covalent bonds, resulting in the migration of hydrogen atoms, e.g. in the histidine side chain, the tautomers have the hydrogen either on Nδ or on Nε as shown.

Transition state: When the energy of the reagents is plotted as the reaction proceeds, the transition state occurs at the point of highest energy.

Van der Waals interactions: A general term to describe *non-bonded* interactions between atoms, comprising dispersion forces, electrostatic interactions, and steric repulsion.

Wild-type: The unmutated form of a gene or genome.

Index

α helix 11, 12
  dipole interactions 76–7
α-lytic protease 45, 73–4
Acetamide 2
Acetyl-CoA 50–2
Acyl enzyme 45–9
Amidase 2
Amino acids
  structures of side chains 7
  properties of side chains 8, 9–11
Aminoacylation 36–8, 69–70
Ampicillin 24
Antibodies 2, 66
  anti-CAMPATH-1 68
  catalytic 68
  reshaping 68
  structure and function 66
  therapy 66

Bacillus stearothermophilus 35–6, 41, 69
B-value 16
Bacterio(phage)
  λ 20–1, 56
  λ promoters 21
  λ repressor 20–1
  M13 36
  T4 DNA ligase 22
  T4 lysozyme 74
β barrel 11, 13
β sheet 11
β lactamase 23
β-galactosidase 20
Binding energy 29, 32
  TyrRS 38–44, 69
Butyramide 2

CAMPATH-1 68
Carboxypeptidase A 35

Catalytic triad 48
Charge relay 46
Chloramphenicol 50–2
Chloramphenicol acetyl transferase 50–2
Chymotrypsin 46, 48, 73
Colony-blot 27
Conformational entropy 76
Crosslinks 77

Denaturation 31
Deoxyuridine triphosphatase 27
Diabetes 61–6
Di-isopropyl fluorophosphate 44
Dipole 9, 41–4, 50, 76–7, 81
Dissociation constant 28–30
Disulphides 10, 48, 61, 74, 75, 77
DNA polymerase 25
DNA ligase 22, 23

EcoR1 23
Electron density 12–17
Electrophile 11, 81
Enthalphy 75, 81
Entropy 24, 44, 74, 76, 77, 81
Enzyme kinetics 29–33
Escherichia coli
  3, 19–24, 36, 53, 56, 62, 69
Eukaryote 19, 48, 81
Evolution 1, 2, 19, 39, 71
Expression 2
  control of 55
  insulin 63
  myoglobin 53
  serine proteases 47–8
  systems 19–22, 23
  TyrRS 36

Filter-binding assay 36
Folding 69, 74
Fourier transform 14
Framework 66–7
Free energy 29–33

Haemoglobin 52
Heavy atom derivatives 14
Helix-turn-helix 56–9
Heteroduplex 25, 27
Hybridization 26–7
Hydrogen bonds 81
  in structures 8–12, 63, 74
  in catalysis
    36–41, 46, 49, 51, 53, 55, 69
  in specificity 39, 49, 51, 53
Hydrogen peroxide 71
Hydrophilic 81
Hydrophobic interactions
  8, 11, 45, 52, 55, 63, 73, 74, 81
Hyperglycaemia 62
Hypervariable loops 67, 68
Hypoglycaemia 62

Inducer 20, 21
Insulin 61–6
  absorption 62
  monomeric 63–6
  therapy 61
Ion-pair 11, 81
Ionic strength 71, 72

k_{cat} 29–31, 36–8, 48–9, 51, 69–70
k_{cat}/K_M 31–3
  in TyrRS 38–40, 69–71
  in serine proteases 48–9, 72
K_M 29–31, 38, 49, 69–71

Lac system 20
  operon 20
  repressor 20, 22
  operator 20
  promoter 53
λ repressor 20
  promoter 21
  operators 20
Leaving group 44, 45, 47
Lysozyme, thermostability 74–8

M-value 55
Main chain 5–8, 11, 49, 76
Melting temperature (Tm) 75
Methionine, oxidation 71
Michaelis–Menten equation
  26, 31–2, 38
Michaelis complex 45
Mistranslation 49
Modelling 14–17
Mutagenesis, directed 3, 25–7
Myoglobin 14, 52–5
  autoxidation 52–5
  discrimination in ligand binding
    52–5
  expression 24
  structure 5, 6, 11

Neutron diffraction 53
NMR 12, 46
Non-Hodgkin lymphoma 68
NsAAPXpNA 48, 73
Nucleophile
  10, 11, 42–4, 45–7, 49, 51, 55, 81

Operator 20, 56, 59
Oxidative stability 71
Oxyanion 49, 51

P_1 pocket 73
Pseudomonas aeruginosa 2
Peptide bond 5, 8–11, 76, 77
Peptide paranitroanilides 48, 73
pH/activity profile 46, 72
Phenylacetamide 2
pK_a 8, 10, 11, 46, 72
Plasmids 23, 25, 27
PMSF 49
Porphyrin 52, 53
Prokaryote 81
Protein synthesis
  5, 36, 42, 47–8, 49, 50, 61

R-factor 15, 16
Ramachandran plot 8, 10, 76
RecA 21
Recognition helices 56–8
  434 56–9
  P22 56–9
  434R[αP22] 56–9
  swapping 57
Repressor 20–2, 24, 55–9

Resolution 12, 15, 16, 52, 63, 74
Restriction endonuclease 22, 23
RNA polymerase 19, 21

Salmonella typhimurium 56
Scissile bond 45–6, 81
Screening 25–7
Serine protease 31, 44–50, 51, 71–4
Side chains
  structures 7
  properties 8, 9–11
Specificity
  29, 36–42, 45–7, 55–60, 73–4
Subtilisin 45, 48, 71, 74

T4 lysozyme 74–8
Tautomer 11, 51, 82
Thermodynamic pit 70
Thiol sepharose 48
TPCK 46
Transcription 19–22, 56
Transformation 24

Translation 19, 41, 49
Transition state
  31–3, 37–9, 41, 44, 47–50, 51, 68, 69–71, 82
Triose phosphate isomerase 11, 13
Trypsin 45, 48
Tyrosyl adenylate 35–44, 69–71
Tyrosyl tRNA synthetase
  35–44, 69–71

Unfolding 69, 74–8
Uracil glycosylase 27

van der Waals 37, 63–4, 82

Wild type
  2, 17, 25–7, 28–9, 31–3, 38–44, 44–50, 51, 67, 71, 73, 77, 82

X-ray crystallography 12–17